Child Maltreatment Solutions Network

This book series is based on an annual conference held by Penn State University's Network on Child Protection and Well-Being. The conference focuses on key issues in child maltreatment. It brings together 200 scholars and policy experts with the following aims:

- Promote interdisciplinary dialogue and stimulate research on child maltreatment
- Promote scholarly excellence by inviting leaders in the field to present and dialogue
- Identify promising issues and solutions in the field of child maltreatment that are not covered by professional societies, foundations, or government agencies

The conference and book series continue the Network's efforts to target a range of issues pertaining to child maltreatment. Approximately ten to fifteen top scholars in the field of child maltreatment convene to present and critique research on the identified topic and to consider the implications for future research and next directions. The conference brings together scholars from diverse fields including child and human development, psychology, education, law, and policy. The symposium organizers, Jennie Noll and Sandee Kyler, also make an effort to include international scholars.

Christian M. Connell • Daniel Max Crowley
Editors

Strengthening Child Safety and Well-Being Through Integrated Data Solutions

Springer

Editors
Christian M. Connell
Human Development and Family Studies
Pennsylvania State University
University Park, PA, USA

Daniel Max Crowley
Human Development and Family Studies
Pennsylvania State University
University Park, PA, USA

ISSN 2509-7156 ISSN 2509-7164 (electronic)
Child Maltreatment Solutions Network
ISBN 978-3-031-36607-9 ISBN 978-3-031-36608-6 (eBook)
https://doi.org/10.1007/978-3-031-36608-6

This Springer imprint is published by the registered company Springer Nature Switzerland AG
The registered company address is: Gewerbestrasse 11, 6330 Cham, Switzerland

Acknowledgments and Dedication

We want to express our thanks and appreciation to the individuals who shared their knowledge and expertise at the Child Maltreatment Solutions Network conference on "Strengthening Child Safety and Wellbeing through Integrated Data Solutions," and particularly to those individuals who contributed chapters to this volume based on that conference. We greatly appreciate your willingness to share your knowledge, as well as your patience as this volume took shape.

In addition, we want to extend our thanks to members of the Child Maltreatment Solutions Network and the Penn State University community who assisted in putting together both the conference and the volume, including: Jennie Noll, Sarah Font, Sheridan Miyamoto, Cheri McConnell, Sandee Kyler, Ashley Stauffer, and Mary McCauley. Your support in planning the conference event helped to create a tremendously informative and insightful conference, contributing to what we feel is an equally informative and insightful volume.

A successful event would not have been possible without the generous support of our conference sponsors including the Social Science Research Institute; Bennett Pierce Prevention Research Center; Institute for CyberScience; Clearinghouse for Military Family Readiness; Child Study Center; College of Information Sciences and Technology; College of Nursing; Department of Biobehavioral Health; Department of Educational Psychology, Counseling, and Special Education; Public Health Services; and the Paterno Library.

We want to dedicate this volume to the individuals who work on behalf of children, youth, and families to ensure that children are raised in environments that support their growth and development and that they are able to experience safety and wellbeing. Finally, we want to thank our families for their love and support throughout the preparation of this volume: Jennifer, Owen, and Clare (CMC); and Nikki, Harrison, and Alice (DMC).

The Pennsylvania State University
University Park, PA, USA

Christian M. Connell
Daniel Max Crowley

Contents

About the Editors

Christian M. Connell, PhD, is an Associate Professor of Human Development and Family Studies at the Pennsylvania State University. Dr. Connell is the Director of the Child Maltreatment Solutions Network, a unit of the university's Social Science Research Institute that engages in novel, interdisciplinary, and translational research to prevent, detect, and treat child abuse and neglect; provides educational programming in child maltreatment and advocacy at the undergraduate, graduate, and post-graduate levels; and serves as the university clearinghouse for child maltreatment information, awareness, and communications. His research focuses on individual, family, and system-level outcomes of youth who have been maltreated or involved in the child welfare and other child-serving systems, as well as the effectiveness of community-based interventions to reduce adverse outcomes associated with maltreatment and trauma. Dr. Connell's research has been funded by the National Institutes of Health, the Children's Bureau of the Administration for Children and Families, and the National Child Traumatic Stress Network of the Substance Abuse and Mental Health Services Administration, and through state and local contracts.

Daniel Max Crowley, PhD, is a Professor of Human Development, Family Studies, and Public Policy and holds the C. Eugene Bennett Endowed Chair in Prevention Research at Pennsylvania State University. Dr. Crowley is the Director of the Edna Bennett Pierce Prevention Research Center. He also directs the Evidence-to-Impact Collaborative, including its infrastructure, the Administrative Data Accelerator. Dr. Crowley's research focuses on improving the use of research in policymaking and optimizing investments in evidence-based interventions. This includes efforts to map the impact of public investments onto public systems through the use of administrative data. Dr. Crowley is the Principal Investigator on grants from the National Institute on Drug Abuse, National Institute of Child Health and Human Development as well as the Annie E. Casey, Laura & John Arnold, Robert Wood Johnson, William T. Grant, and Doris Duke Charitable Foundations. Dr. Crowley has received national awards recognizing his scholarship from the National Institutes of Health, National Bureau of Economic Research, Society for Prevention Research, Association for Public Policy & Management, Research Society on Alcoholism, and National Prevention Science Coalition.

About the Authors

Lawrence M. Berger is Associate Vice Chancellor for Research in the Social Sciences, Vilas Distinguished Achievement Professor in the Sandra Rosenbaum School of Social Work, and past Director of the Institute for Research on Poverty at the University of Wisconsin-Madison. His research focuses on the ways in which economic resources, sociodemographic characteristics, and public policies affect parental behaviors and child and family wellbeing. He is engaged in studies in three primary areas: (1) examining the determinants of substandard parenting, child maltreatment, and out-of-home placement for children; (2) exploring associations among socioeconomic factors (family structure and composition, economic resources, household debt), parenting behaviors, and children's care, development, and wellbeing; and (3) assessing the influence of public policies on parental behaviors and child and family wellbeing. His work aims to inform public policy in order to improve its capacity to assist families in accessing resources, improving family functioning and wellbeing, and ensuring that children are able to grow and develop in the best possible environments.

Stephanie Cuccaro-Alamin, PhD, is a Research Specialist at the California Child Welfare Indicators Project (CCWIP) and a Researcher at the Children's Data Network (CDN). Her research interests include child welfare services, poverty, risk assessment, and the use of linked administrative data for program evaluation and policy research. Stephanie graduated with a BS in Psychology from Union College (NY), received her MSW from Columbia University, and earned her PhD in Social Welfare from the University of California at Berkeley.

Brett Drake, PhD, is a Professor of Data Science for the Social Good in Practice at the Brown School of Social Work at Washington University in St. Louis. His substantive area is child maltreatment and public child welfare systems, with an emphasis on early system contacts, including reporting and substantiation. Prior to joining academia, Dr. Drake worked in Child Protective Services for 3 years.

Bridget Freisthler, PhD, is a Professor in the College of Social Work at The Ohio State University. Her research focuses on how situation, social, and location characteristics affect parenting. She is interested in how microcontexts (e.g., where a person is during a particular event) shape behavior, such as parenting or substance use. These microcontexts might include where and with whom parents drink alcohol, use marijuana, or experience higher levels of stress; all of these may affect parenting behaviors.

Melissa Jonson-Reid, PhD, is the Ralph and Muriel Pumphrey Professor of Social Work at Washington University's George Warren Brown School of Social Work, Associate Dean of Transdisciplinary Faculty Affairs, and Fellow of the American Academy of Social Work and Social Welfare. She practiced as a school social worker and project director of a program called Foster Youth Services in two California school districts that served youth in child welfare and probation care prior which led to her interests in children's services trajectories and policy. Her work has two primary foci: (1) administrative data research on cross-sector service response for low income and maltreated populations and to assess services in collaboration with community agencies to look for opportunities to improve services and policy to improve longer term outcomes; and the (2) second focal area is child maltreatment prevention.

Nancy Jo Kepple, PhD, MSW, is an Associate Professor at the School of Social Welfare, University of Kansas and an Affiliate Faculty at the Cofrin Logan Center for Addiction Research and Treatment. She earned her MSW and PhD from the University of California, Los Angeles. Dr. Kepple's research broadly examines the social consequences of the availability, distribution, and use of psychoactive substances. Her current research focuses on the interplay between parent substance use, neuropsychological functioning, and social environment and their relationship to parenting behaviors.

Jennie G. Noll, PhD, is the Ken Young Family Professor for Healthy Children at the Department of Human Development and Family Studies, College of Health and Human Development, The Pennsylvania State University. She directs the Center for Safe and Healthy Children, is the Principal Investigator (PI) of the NICHD P50 Capstone Center for Excellence, the Translational Center for Child Maltreatment Studies, and MPI of the T32 training grant Creating the Next Generation of Scholars in the Child Maltreatment Sciences. Through contiguous R-level NIH funding over the past three decades, Dr. Noll has been the PI of several prospective, longitudinal cohort studies of the impact of abuse and neglect throughout development and across generations. Her primary research foci include: the bio-psycho-social consequences of childhood sexual abuse, pathways to teen pregnancy and high-risk sexual behaviors for abused and neglected youth, the long-term adverse health outcomes of abuse survivors, midlife reversibility of neurocognitive deficits in stress-exposed populations, the impact of high-risk internet and social media behaviors on teen development, and the primary prevention of sexual abuse. The chief thrust of

Dr. Noll's research and infrastructure grants is to leverage cutting-edge science to aid evidence-informed policymaking that implores a larger public investment in the prevention and treatment of child abuse and neglect.

María Gandarilla Ocampo, MSW, is a social work doctoral candidate at the Brown School at Washington University in St. Louis studying child maltreatment prevention and child maltreatment mandated reporting policies. Her work mainly focuses on the implementation of child maltreatment mandated reporting policies and the impact they have on families and child welfare systems. María hopes that her work can inform the development of equitable policies and community-based intervention programs that support families to be well and thrive in their communities.

Emily Putnam-Hornstein, PhD, is the John A. Tate Distinguished Professor for Children in Need at the University of North Carolina at Chapel Hill. She also maintains appointments as a Distinguished Scholar at the University of Southern California where she co-directs the Children's Data Network and as a Research Specialist with the California Child Welfare Indicators Project at UC Berkeley. Her analysis of large-scale, linked administrative data has provided insight into where scarce resources may be most effectively targeted and informs understanding of maltreated children within a broader, population-based context. Her research has been used to develop risk stratification tools, including those implemented in Allegheny County, Pennsylvania, and Los Angeles County, California. These tools support caseworkers and supervisors in reviewing hundreds of factors relevant to a child's risk and safety when making initial screening and triaging decisions. Emily is the recipient of the Forsythe Award for Child Welfare Leadership from the National Association of Public Child Welfare Administrators and the Commissioner's Award from the Children's Bureau. Emily graduated from Yale University with a BA in Psychology, received her MSW from Columbia University, and earned her PhD in Social Welfare from the University of California at Berkeley.

Rhema Vaithianathan is a Professor of Health Economics at Auckland University of Technology (AUT), New Zealand. She specializes in the use of integrated administrative data with machine learning tools in health and human services including child welfare. Her work is focused on translational research and she led the development and deployment of the Allegheny Family Screening Tool, one of the first uses of predictive risk models in Child Welfare.

Elizabeth Weigensberg, PhD, is a Principal Researcher at Mathematica. She has expertise in designing and conducting evaluations, using both qualitative and quantitative methods. Her expertise includes linking and analyzing complex administrative data from state and local public agencies and providing technical assistance to facilitate the development and use of data to inform policy and practice. Dr. Weigensberg directs several projects for the U.S. Department of Health and Human Services, including the State Child Abuse and Neglect (SCAN) Policies Database and the Child Welfare Study to Enhance Equity with Data (CW-SEED) for the Administration for Children and Families, Office of Planning, Research, and Evaluation. She is also the Project Director for the Child Welfare and Health Infrastructure for Linking and Data Analysis of Resources, Effectiveness, and Needs (CHILDREN) Initiative for the Office of the Assistant Secretary for Planning and Evaluation, which supports states with developing their linked data infrastructure and analytic capacity with State child welfare and Medicaid data. Before Dr. Weigensberg came to Mathematica in 2015, she previously worked as a senior researcher at Chapin Hall at the University of Chicago, as a Research Instructor at the University of North Carolina at Chapel Hill, and as an Analyst at the U.S. Government Accountability Office. She holds a PhD in Social Work from the University of North Carolina at Chapel Hill and an MS in Social Work from Columbia University.

Christopher Wildeman, PhD, is a Professor of Sociology in the Trinity College of Arts and Sciences at Duke University, where he is also Director of the National Data Archive on Child Abuse and Neglect (NDACAN). Since 2019, he has also been Research Professor at the ROCKWOOL Foundation Research Unit in Copenhagen, Denmark. His research interests center around estimating the prevalence, causes, and consequences of contact with the child welfare system and the criminal legal system.

Jennifer Price Wolf, PhD, MPH, MSW, is an Associate Professor at the School of Social Work, San Jose State University. Price Wolf's research examines intersections between public health and social work, including how community environments influence substance use behaviors and family well-being.

Youngmin Yi, PhD, is an Assistant Professor of Sociology at the University of Massachusetts Amherst. Her research focuses on the intersection of family life with the criminal legal, child welfare systems in the United States as well as critical approaches to the use and analysis of quantitative data for the study of social inequality in family life.

Chapter 1
Introduction and Volume Overview

Christian M. Connell

Child maltreatment—including physical, sexual, and emotional abuse, and neglect—is a significant public health problem resulting in substantial adverse consequences for children and families and for society at large. Individual costs are reflected in the adverse short- and long-term physical or psychological effects of maltreatment on victims. Societal costs associated with child maltreatment include increased health care, criminal justice, and educational and economic consequences, and are estimated at an annual US population lifetime cost of $124 billion [2010 USD] for substantiated victims and $585 billion for investigated cases (Fang et al., 2012), with more recent estimates rising even higher ($428 billion and $2.0 trillion [2015 USD], respectively; Peterson et al., 2018). Yet, federal, state, and local governments face substantial barriers in the identification and assessment of maltreatment and in providing intervention and treatment services. The scope and complexity of child maltreatment, coupled with the limited resources available to the child welfare system, underscore the need for programmatic and policy-level solutions that are demonstrably effective and efficient in promoting child safety, permanency, and well-being.

Over the past two decades, the landscape for using data to inform research and policy development efforts affecting child protective services (CPS) and child welfare system (CWS) activities and outcomes has seen tremendous growth. In particular, the field has witnessed a significant rise in the use of administrative data resources—and increasingly linked or integrated administrative data systems—to carry out this critical work. At this point, the use of administrative data is a relatively common aspect of research activities related to child maltreatment and child safety concerns. A cursory search of two of the prominent scientific journals in the

C. M. Connell (✉)
Human Development and Family Studies & Child Maltreatment Solutions Network, Pennsylvania State University, University Park, PA, USA
e-mail: cmc128@psu.edu

C. M. Connell, D. M. Crowley (eds.), *Strengthening Child Safety and Well-Being Through Integrated Data Solutions*, Child Maltreatment Solutions Network, https://doi.org/10.1007/978-3-031-36608-6_1

discipline reveals that between 2018 and 2022, *Child Maltreatment* published a total of 63 research articles that included the term "administrative data" or related terms (e.g., CPS records, child maltreatment records, SACWIS); *Child Abuse & Neglect* published 196 research articles including those terms in the same period.

Administrative data refers to data collected by public agencies to support normal practice operations (e.g., for record keeping, case management, billing, or program/performance monitoring) and typically maintained in some form of electronic record (DeHart & Shapiro, 2017; Jonson-Reid & Drake, 2008). In the child maltreatment research field, administrative data most frequently include CPS-related records such as child-, family-, or case-level information pertaining to individuals involved in a CPS allegation. These data often include demographic information (e.g., dates of birth, gender, race, and ethnicity), case-related details (e.g., types of alleged maltreatment, dates of reporting), case findings, and system response (e.g., provision of services, out-of-home placement; Hurren et al., 2017). They may also include information about child welfare system placement such as key dates (e.g., dates of entry or removal, placement change, discharge), types of service (e.g., kinship or non-relative foster home, group home, shelter), or placement outcomes (e.g., exiting to reunification or adoption). Other public systems (e.g., education, health care, criminal justice, or public welfare programs) generate comparable administrative records to manage operations or track program involvement and outcomes. Health care records, such as state Medicaid programs, may include details on program beneficiaries, diagnostic information, service contacts and procedures, access to prescriptions, or use of inpatient facilities. Criminal justice system data may include details related to arrests (e.g., dates, types of charges), court outcomes (e.g., adjudication), and subsequent sentencing or sanction decisions (e.g., placement, probation, or parole). Data are typically organized at multiple levels such that identifiers allow for identification of individual system contacts (e.g., separate CPS reports or foster care placement episodes, individual health care encounters), as well as the tracking of individuals and/or families involved in a given system over-time. Data may also include identifiers for individual service providers or caseworkers who come into contact with a given case.

Though such data systems are developed to support program operations and administrative functions, they may be leveraged by researchers to create cross-sectional or longitudinal cohorts to study outcomes within a given system. The use of administrative data for research purposes offers a number of important advantages over other methods of data collection with respect to child maltreatment or child welfare systems research. Frequently cited benefits include the availability of large and comprehensive datasets covering the total population of individuals served by a given agency for specific periods of time, sufficient numbers of individuals to facilitate analyses of small subgroups and support complex and multi-level analyses, and data that are less subject to recall bias or social desirability than other means of data collection (Brownell & Jutte, 2013; Glasson & Hussain, 2008; Hurren et al., 2017; Yampolskaya, 2017). Administrative data may also be more time- and cost-efficient to obtain if systems have sufficient infrastructure in place to support access to and extraction of these data (Penner & Dodge, 2019). Finally, inclusion of variables or data elements that are relevant to agency and program monitoring has been cited as a strength, in that such data lend themselves to answering research questions relevant to decision and policy makers in these settings.

Despite these benefits, administrative data systems also pose a number of potential limitations that researchers must be aware of (Hurren et al., 2017). First, while datasets may represent the full population of individuals served by a particular agency (e.g., CPS), they do not reflect the full scope of individuals who experience maltreatment, since many such individuals may never come to the attention of CPS. In addition, researchers often have limited input into the types of variables included in administrative datasets, and such data were not necessarily intended to support research activities or measure constructs of interest to researchers with sufficient psychometric properties. Thus, researchers may need to invest significant time and attention to understanding the nature of data elements (e.g., how they are operationalized, what factors contribute to their collection or inclusion, and what factors may contribute to their missingness in a data extract). Relatedly, researchers often have to invest considerable time and resources in managing and coding data extracts to prepare them for data analytic phases of research. Finally, it can take time to secure data agreements necessary to obtain these data, including negotiation of various legal concerns and data security requirements related to their transfer, storage, and access—and these processes may also be affected by changes in agency leadership or priorities.

Single-system administrative data extracts (e.g., such as those drawn from a single state or county CPS agency) may be used to link files involving the same individuals or families over time to create a longitudinal cohort to investigate system-level outcomes. Child maltreatment research in this vein may involve construction of a longitudinal statewide cohort of youth with a history of involvement in CPS allegations. Such a cohort may be used to investigate differences in patterns of system contact based on child or family characteristics (Fix & Nair, 2020; Jud et al., 2016), factors associated with risk of re-report (Connell et al., 2007; Jonson-Reid et al., 2003; Kohl & Barth, 2007; Putnam-Hornstein et al., 2015), or factors associated with recurrent substantiated cases of abuse or neglect (Connell et al., 2009; Fluke et al., 2008; Holbrook & Hudziak, 2020; Kim & Drake, 2019). Data from multiple states (e.g., using extracts from multiple states or using Federal data sources such as the National Child Abuse and Neglect Data System [NCANDS]) could be used to conduct comparative analyses of such outcomes across different community settings (e.g., across county or state systems; Fluke et al., 1999; Johnson-Motoyama et al., 2022; Maguire-Jack et al., 2020). Given the types of data elements typically available in child welfare administrative data systems, research using single-system administrative data extracts usually focus on safety-related outcomes (e.g., types of maltreatment, risks of recurrence, recidivism among perpetrators). Fewer studies using child welfare system data, in insolation, have access to indicators of child well-being (e.g., child-level assessments of physical, behavioral, or mental health or functioning).

To expand the range of outcomes and research questions available using administrative data, researchers have increasingly been turning to the use of integrated administrative data systems, which involve a variety of methods to link data involving extracts from two or more distinct agency administrative data systems. Linking involves the tracking of individuals or families across different administrative data systems. Linking may be carried out through deterministic methods when data systems share common identifier fields that serves as a consistent and error-free means of tracking individuals across systems (Shlomo, 2019). Some states, for example, have developed universal identifiers that may be included in multiple state administrative data systems (e.g., CPS, Medicaid, entitlement programs). In the absence of a consistent identifier, linking may be accomplished through probabilistic methods in which multiple fields (including names, dates of birth, personal characteristics like sex, or shared identifier fields like Social Security Number) are compared and researchers set a threshold to identify individuals across systems (Christen, 2012; Harron et al., 2015; Herzog et al., 2007). In practice, these methods are often combined, with deterministic matches being used to locate exact matches, followed by the use of probabilistic methods to identify additional individuals in common across systems (Shlomo, 2019).

Integration of administrative data systems offers a number of advantages over traditional single-system administrative data systems. Among the most central benefits is the opportunity to incorporate information from additional domains to provide a more comprehensive understanding of factors that contribute to child safety or well-being or to reflect a broader range of outcomes than would typically be available within a single-agency administrative data source (Jonson-Reid & Drake, 2008). Through data linkage across CPS and health care records, for example, researchers may be able to incorporate indicators of child physical or mental health through diagnostic or assessment data (e.g., Putnam-Hornstein et al., 2022; Rebbe et al., 2019). Integration of educational records may provide information on academic or standardized test performance (e.g., Fantuzzo et al., 2011; Ryan et al., 2018), and linkages with community service provider systems might be able to provide a broader picture of service access or utilization (e.g., Garcia et al., 2018). In addition, access to linked data also provides a means of assessing overlap among service populations (e.g., the relation between CWS and public entitlement programs; Cancian et al., 2013) or cross-over from one system to another (e.g., to examine risks associated with dual system-involved youth; Ryan et al., 2007; Vidal et al., 2017).

Many of the challenges inherent to administrative data are also applicable to the use of linked administrative data systems. Researches may still lack input on inclusion of data elements and must understand the unique factors that contribute to inclusion of individuals or specific indicators and measures in each of the separate administrative organizations. Other challenges can be compounded by the inclusion of multiple separate organizations. There may be additional policy hurdles in place that limit data sharing across organizational units or require the development of multiple data use agreements, each requiring separate reviews or legal consultations prior to initiation of any data sharing. Lack of common identifiers or variability in data quality for shared indicators, such as those used for probabilistic data matching, may also raise concerns about the potential biasing effects of false matches, or of missed matches, on subsequent analyses. Thus, before realizing the time- or cost-efficiencies in research involving integrated administrative data systems, there is a need for significant investment of time and resources to develop the infrastructure to support access, interoperability, and linkage of data systems.

A recent scoping review article identified 121 research articles within the child maltreatment literature that included linked administrative data records from at least two distinct data-holding organizations (Soneson et al., 2022). The review highlighted the growth in use of linked administrative data over the past two decades of child maltreatment research: of the 121 research articles identified, 11 were published prior to 2005, another 11 were published in the next five-year span (2006–2010), 33 in the next five years (2011–2015), and 66 from 2016 to 2020. Nearly all studies involved used population-based linked files (85%), though about 16 percent included linkage to study-specific research data (e.g., surveys or similar). The most common data systems involved included social service data (91% of studies, with CPS records involved in 79% of all studies), followed by health records (76% of studies), justice records (23% of studies), and education (21% of studies). The vast majority of studies reviewed used linked files for descriptive purposes (91% of studies), with relatively fewer published studies conducted to evaluate interventions or services (7%), advance research methods (5%), or support predictive modeling (3%). The authors also noted that few of the studies provided sufficient detail on the quality of data linkage, despite most employing some level of probabilistic matching (63% of studies). As is evident from this review, the field of integrated administrative data systems research is still in its early stages, and there are opportunities to expand the range of systems brought into this area of research, as well as the types of research questions to which such methods may be applied.

1.1 The Child Maltreatment Solutions Network Conference: Strengthening Child Safety and Well-Being Through Integrated Data Solutions

To highlight these issues, as well as the promise and challenge of using integrated administrative data to advance research on child safety and well-being, the Child Maltreatment Solutions Network at the Pennsylvania State University hosted a two-day conference titled “Strengthening Child Safety and Wellbeing through Integrated Data Solutions” in the fall of 2018 on the Penn State University Park campus. The conference brought together a number of leaders in the field of child maltreatment research known for their innovative use of public administrative data systems to study children, families, and communities at risk of child maltreatment or adverse outcomes associated with such experiences. The purpose of the conference was to showcase emerging and innovative approaches in the acquisition and use of administrative data to inform the societal and governmental response to child maltreatment. Broadly, the conference sessions were organized around four sessions addressing substantive content areas relevant to advancing this work (see the appendix in this volume for a list of sessions and speakers). The first session focused on the use of administrative data sources to better understand the scope and impact of child maltreatment at a population level. Session two focused on use of a variety of integrated data sources to identify effective services and interventions appropriate for a child welfare context. The third session shifted the focus to research employing administrative data systems to inform public policy or support system reform efforts. The conference culminated with a final session that included two separate panel discussions to assist with priority setting and active planning for a national agenda on the use of integrated data systems to improve child safety and well-being in the context of risks of child maltreatment or child welfare system involvement.

The chapters in this volume provide compelling examples of the ways in which leaders in the field of child maltreatment research leverage administrative data and integrate with other data sources to investigate issues associated with child safety and well-being. Each of the chapters describes innovative means to advance the field's understanding of child maltreatment and its effects through the use of linked administrative data sources. The aim of this volume is to promote sounder research practices, foster the use of integrated administrative data to investigate risk and protective processes associated with child maltreatment, and expand the breadth of outcomes that can be investigated using such methods. Details on chapter content are summarized below.

Youngmin Yi and Christopher Wildeman discuss the use of synthetic cohort life table methods to estimate prevalence and timing of child welfare system involvement or to examine factors that contribute to risk of involvement or system-level outcomes. Critically, the authors demonstrate the use of these methods with publicly available child welfare data sources (i.e., the National Child Abuse and Neglect Data System (NCANDS) and Adoption and Foster Care Analysis and Reporting System (AFCARS) datasets)—providing an excellent introduction to the use of administrative data for research purposes given ready access to these resources. These data offer a number of advantages relative to more specialized administrative child welfare data sources, including child welfare population representation for all US states, a standardized format consistent across time and locale, the ability to follow cases over multiple submission years, and the ability to link with related Federal datasets through common identifier fields shared among NCANDS, AFCARS, and National Youth in Transition Dataset (NYTD) files. The authors also offer potential solutions to methodological challenges inherent in nationally focused research leveraging these data, such as the difficulty of accounting for movement across states resulting from the lack of a consistent identifier that follows youth or families marked by such mobility.

Emily Putnam-Hornstein, Stephanie Cuccaro-Alamin, and Rhema Vaithianathan present the integration of vital birth records with child welfare system data as an under-utilized resource for conducting population-level analyses of child maltreatment health and safety outcomes. The universality of these records, as well as the critical information contained in standardized birth records files, has particular utility for creating a population registry that can be linked to other core administrative data sources to conduct prospective research. The authors draw on their experiences with the Children's Data Network (CDN) in California to describe numerous applications that use population-level cohorts based on birth records data to investigate risks of maltreatment, intergenerational patterns of abuse and neglect, child death and other rare events, or to monitor community-level risks and assets to inform sound policies and programs for children and families.

Integration of spatial data with CPS records through geocoding of address and location information is the focus of the chapter by Bridget Freisthler, Nancy Kepple, and Jennifer Wolf. The authors discuss advances in this approach that move beyond a child's or family's residential neighborhood to a focus on "activity spaces"—the community settings a parent or child may move through in the course of their daily life such as work, school, or other such settings. This broader perspective on the role of community context provides a richer understanding of the types of environmental exposures to both risks and assets that may affect a child's risk of experiencing child maltreatment or related health and safety concerns. While some research on these types of spatial-contextual factors employ novel data collection methods (e.g., so-called "wearables" such as GPS-enabled watches or other trackers), the authors also explore methods by which integrated data systems can provide valuable information about critical activity spaces, as well as the challenges researchers may experience in constructing such information.

Next, Lawrence Berger provides an overview of a longstanding state-level collaboration utilizing integrated data systems that link social welfare program administrative data systems (e.g., SNAP, TANF, unemployment insurance) with health service, child welfare (including CPS and foster care placement), incarceration, and public-school records. The chapter provides a detailed history of the development of the Wisconsin Administrative Data Core resource—including key partnerships and funding—and describes numerous applications of this integrated database to address complex issues for children and families coming into contact with these various systems. The chapter highlights the immense value of integrating public administrative data records across state systems to inform practice and policy matters related to child safety and well-being, while also demonstrating the significant commitment of multiple partners to carrying out such work.

Melissa Jonson-Reid, Brett Drake, and Maria Ocampo use the astronomical concept of the "Goldilocks Zone"—the sweet spot where conditions are such that life may flourish—to discuss the application of integrated data to promote greater understanding of appropriate levels of service utilization among families involved with child welfare, as well as the optimization of resources within the child welfare system. Linking administrative data systems that include service and resource utilization data provides a means of understanding patterns of service use at a broader system level, as well as factors that contribute to or inhibit access to resources. These types of integrated data systems also provide a way for researchers to study "treatment as usual" within a child welfare context and to evaluate the effectiveness of new programs or policies relative to standard care practices. The authors caution against potential problems or concerns that may be encountered with access to linked data, including careful consideration of data security and the need to avoid drawing conclusions when data quality may be a concern.

Elizabeth Weigensberg continues the focus on use of integrated data systems to investigate service utilization patterns. Her chapter describes a multi-state study, conducted by Mathematica and other collaborators, that linked state child welfare and Medicaid records to identify patterns of "super-utilization" (i.e., notably high patterns of service use or cost) of child welfare and health care services among children in foster care in two states—Tennessee and Florida. Integration of child welfare and health care records allowed the study to investigate patterns of service use intensity, duration, frequency, and cost that could be used to identify individuals at the upper ends of service-utilization within these two systems, as well as to examine the overlap of these two populations through a variety of advanced statistical methods. Such information may be invaluable to strengthening coordination of services across systems to reduce costs and provide services more efficiently.

In a second contribution to this volume, Rhema Vaithianathan, Stephanie Cuccaro-Alamin, and Emily Putnam-Hornstein describe the use of predictive risk modeling (PRM) methodologies to motivate changes in practice and policy at multiple stages of the child welfare system response. PRM leverages integrated administrative data and "big data" analytic strategies to advance the science of child welfare decision-making beyond those of more traditional methods (e.g., such as those using actuarial or consensus-based models of risk). The authors offer a number of real-world case studies, drawn from their experiences in both the USA and New Zealand, to demonstrate how PRM can improve practice and enhance decision-making by caseworkers and other child welfare staff. The authors also discuss a number of critical ethical considerations to ensure that the use of algorithm-based methods support, rather than replace, more effective decision-making within child welfare systems.

The final chapter of this volume presents two panel discussions led by faculty from the Pennsylvania State University Child Maltreatment Solutions Network. The first panel, led by Jennie Noll, featured an open discussion with child welfare administrators and other officials representing numerous county and state systems in Pennsylvania and other parts of the USA. Participants raised important questions about the use of integrated and administrative data systems, as well as highlighted applications, or potential applications, from their own settings. The second panel, led by Christian Connell, reconvened conference presenters for a broader reflection on the promise and challenge of using integrated data systems to advance research on child safety and well-being, as well as to highlight areas of future direction for the field to pursue with such tools and resources. These discussions emphasized the importance of building sound partnerships among academic institutions, public agencies, and funders to maximize the returns that may be realized from such efforts. The presenters also reflected on the need for greater guidance and support at the Federal level to overcome barriers—including fiscal and policy-level challenges—and highlighted the types of information that are often lacking in such efforts to more adequately measure and document child and caregiver well-being.

1.2 Conclusion

Ultimately, integrated administrative data systems to conduct research provide another tool for applied researchers to investigate factors contributing to the safety, health, and well-being of children and families. The goals of the conference, and of this resulting volume, are to provide a framework for understanding how such systems might be developed and used to conduct this important research to expand our understanding of risks of child maltreatment and its adverse consequences for health and well-being, as well as to strengthen the infrastructure for conducting research on services, supports, and interventions to prevent child maltreatment or to ameliorate negative effects. Finally, consistent with the mission and aims of the Pennsylvania Child Maltreatment Solutions Network, we want to inform the ways in which these types of research methods and tools can be translated to inform the policies and practices of the broader child welfare system and efforts to promote child safety and well-being.

References

Brownell, M. D., & Jutte, D. P. (2013). Administrative data linkage as a tool for child maltreatment research. *Child Abuse & Neglect, 37*(2–3), 120–124.

Cancian, M., Yang, M., & Slack, K. S. (2013). The effect of additional child support income on the risk of child maltreatment. *Social Service Review, 87*(3), 417–437.

Christen, P. (2012). *Data matching: Concepts and techniques for record linkage, entity resolution, and duplicate detection*. Springer.

Connell, C. M., Bergeron, N., Katz, K. H., Saunders, L., & Tebes, J. K. (2007). Re-referral to Child Protective Services: The influence of child, family, and case characteristics on risk status. *Child Abuse & Neglect, 31*, 573–588.

Connell, C. M., Vanderploeg, J. J., Katz, K. H., Caron, C., Saunders, L., & Tebes, J. K. (2009). Maltreatment following reunification: Predictors of subsequent child protective services contact after children return home. *Child Abuse & Neglect, 33*, 218–228.

DeHart, D., & Shapiro, C. (2017). Integrated administrative data & criminal justice research. *American Journal of Criminal Justice, 42*(2), 255–274.

Fang, X., Brown, D. S., Florence, C. S., & Mercy, J. A. (2012). The economic burden of child maltreatment in the United States and implications for prevention. *Child Abuse & Neglect, 36*(2), 156–165.

Fantuzzo, J. W., Perlman, S. M., & Dobbins, E. K. (2011). Types and timing of child maltreatment and early school success: A population-based investigation. *Children and Youth Services Review, 33*(8), 1404–1411.

Fix, R. L., & Nair, R. (2020). Racial/ethnic and gender disparities in substantiation of child physical and sexual abuse: Influences of caregiver and child characteristics. *Children and Youth Services Review, 116*, 105186.

Fluke, J. D., Yuan, Y. T., & Edwards, M. (1999). Recurrence of maltreatment: An application of the National Child Abuse and Neglect Data System (NCANDS). *Child Abuse & Neglect, 23*(7), 633–650.

Fluke, J. D., Shusterman, G. R., Hollinshead, D. M., & Yuan, Y. T. (2008). Longitudinal analysis of repeated child abuse reporting and victimization: Multistate analysis of associated factors. *Child Maltreatment, 13*(1), 76–88.

Garcia, A. R., Metraux, S., Chen, C.-C., Park, J. M., Culhane, D. P., & Furstenberg, F. F. (2018). Patterns of multisystem service use and school dropout among seventh-, eighth-, and ninth-grade students. *The Journal of Early Adolescence, 38*(8), 1041–1073.

Glasson, E. J., & Hussain, R. (2008). Linked data: Opportunities and challenges in disability research. *Journal of Intellectual and Developmental Disability, 33*(4), 285–291.

Harron, K., Goldstein, H., & Dibben, C. (Eds.). (2015). *Methodological developments in data linkage*. John Wiley & Sons.

Herzog, T. N., Scheuren, F. J., & Winkler, W. E. (2007). *Data quality and record linkage techniques*. Springer Science & Business Media.

Holbrook, H. M., & Hudziak, J. J. (2020). Risk factors that predict longitudinal patterns of substantiated and unsubstantiated maltreatment reports. *Child Abuse & Neglect, 99*, 104279.

Hurren, E., Stewart, A., & Dennison, S. (2017). New methods to address old challenges: The use of administrative data for longitudinal replication studies of child maltreatment. *International Journal of Environmental Research and Public Health, 14*(9), 1066–1077.

Johnson-Motoyama, M., Ginther, D. K., Phillips, R., Beer, O. W., Merkel-Holguin, L., & Fluke, J. (2022). Differential response and the reduction of child maltreatment and foster care services utilization in the US from 2004 to 2017. *Child Maltreatment*, 10775595211065761.

Jonson-Reid, M., & Drake, B. (2008). Multisector longitudinal administrative databases: An indispensable tool for evidence-based policy for maltreated children and their families. *Child Maltreatment, 13*(4), 392–399.

Jonson-Reid, M., Drake, B., Chung, S., & Way, I. (2003). Cross-type recidivism among child maltreatment victims and perpetrators. *Child Abuse & Neglect, 27*(8), 899–917.

Jud, A., Fegert, J. M., & Finkelhor, D. (2016). On the incidence and prevalence of child maltreatment: A research agenda. *Child and Adolescent Psychiatry and Mental Health, 10*(1), 1–5.

Kim, H., & Drake, B. (2019). Cumulative prevalence of onset and recurrence of child maltreatment reports. *Journal of the American Academy of Child & Adolescent Psychiatry, 58*(12), 1175–1183.

Kohl, P. L., & Barth, R. P. (2007). Child maltreatment recurrence among children remaining in-home: Predictors of re-reports. In R. Haskins, F. Wulczyn, & M. B. Webb (Eds.), *Child protection: Using research to improve policy and practice* (pp. 207–225). Brookings Institution Press.

Maguire-Jack, K., Font, S. A., & Dillard, R. (2020). Child protective services decision-making: The role of children's race and county factors. *American Journal of Orthopsychiatry, 90*(1), 48.

Penner, A. M., & Dodge, K. A. (2019). Using administrative data for social science and policy. *RSF: The Russell Sage Foundation Journal of the Social Sciences, 5*(2), 1–18.

Peterson, C., Florence, C., & Klevens, J. (2018). The economic burden of child maltreatment in the United States, 2015. *Child Abuse & Neglect, 86*, 178–183.

Putnam-Hornstein, E., Simon, J. D., Eastman, A. L., & Magruder, J. (2015). Risk of re-reporting among infants who remain at home following alleged maltreatment. *Child Maltreatment, 20*(2), 92–103. https://doi.org/10.1177/1077559514558586

Putnam-Hornstein, E., Foust, R., Cuccaro-Alamin, S., Prindle, J., Nghiem, H., Ahn, E., & Palmer, L. (2022). A population-based study of mental health diagnoses and child protection system involvement among Medicaid-insured children. *The Journal of Pediatrics*.

Rebbe, R., Mienko, J. A., Brown, E., & Rowhani-Rahbar, A. (2019). Hospital variation in child protection reports of substance exposed infants. *The Journal of Pediatrics, 208*, 141–147. e142.

Ryan, J. P., Herz, D., Hernandez, P. M., & Marshall, J. M. (2007). Maltreatment and delinquency: Investigating child welfare bias in juvenile justice processing. *Children and Youth Services Review, 29*(8), 1035–1050.

Ryan, J. P., Jacob, B. A., Gross, M., Perron, B. E., Moore, A., & Ferguson, S. (2018). Early exposure to child maltreatment and academic outcomes. *Child Maltreatment, 23*(4), 365–375.

Shlomo, N. (2019). Overview of data linkage methods for policy design and evaluation. In N. Crato & P. Paruolo (Eds.), *Data-driven policy impact evaluation: How access to microdata is transforming policy design* (pp. 47–65). Springer.

Soneson, E., Das, S., Burn, A.-M., Van Melle, M., Anderson, J. K., Fazel, M., Fonagy, P., Ford, T., Gilbert, R., & Harron, K. (2022). Leveraging administrative data to better understand and address child maltreatment: A scoping review of data linkage studies. *Child Maltreatment*, 10775595221079308.

Vidal, S., Prince, D., Connell, C. M., Caron, C. M., Kaufman, J. S., & Tebes, J. K. (2017). Maltreatment, family environment, and social risk factors: Determinants of the child welfare to juvenile justice transition among maltreated children and adolescents. *Child Abuse & Neglect, 63*, 7–18. https://doi.org/10.1016/j.chiabu.2016.11.013

Yampolskaya, S. (2017). Research at work: Administrative data and behavioral sciences research. *Families in Society, 98*(2), 121–125.

Chapter 2
How the AFCARS and NCANDS Can Provide Insight into Linked Administrative Data

Youngmin Yi and Christopher Wildeman

Scholars, practitioners, and policymakers have long known a great deal about the characteristics, experiences, and later life outcomes of children who come into contact with the child welfare system. However, until recently, more fundamental questions about the prevalence and distribution of child welfare system involvement remained unanswered. *What proportion of children have contact or will ever have contact with the child welfare system? How do these risks differ across jurisdictions, states, and social groups? If there are differences in risks of contact, do they shift across types and levels of system involvement or are they relatively constant?* The ideal set of data and methods to answer these types of questions must meet some key requirements.

First, these data must include information about the entire population, rather than exclusively about the subgroup of children known to be involved or to have been involved with the child welfare system. This requirement is necessary because without knowing the number of children at risk of experiencing an event, it is impossible to estimate the proportion of those children who actually experience the event. Second, these data must observe the entire population of children from birth through age 18. Third, these data must include information about the child welfare events and exposures of interest. Finally, if any level of disaggregation along demographic, social, economic, and other characteristics and groups is of interest, such information would also need to be available. Population registers such as those found in Denmark (e.g. Andersen & Wildeman, 2014; Fallesen, 2014), Sweden (e.g. Vinnerljung et al., 2005), and Norway (Staer & Bjørknes, 2015) meet these criteria

Y. Yi (✉)
University of Massachusetts Amherst, Amherst, MA, USA
e-mail: youngminyi@umass.edu

C. Wildeman
Duke University, Durham, NC, USA

ROCKWOOL Foundation Research Unit, Copenhagen, Denmark

C. M. Connell, D. M. Crowley (eds.), *Strengthening Child Safety and Well-Being Through Integrated Data Solutions*, Child Maltreatment Solutions Network, https://doi.org/10.1007/978-3-031-36608-6_2

within the context of the child welfare system. However, there is currently no such national data system in the USA for the child welfare system.

This chapter proceeds in three parts. First, we illustrate how synthetic cohort life table analysis of the Adoption and Foster Care Analysis and Reporting System Data (AFCARS) and the National Child Abuse and Neglect Data System (NCANDS) has substantially enhanced our knowledge of the reach and unequal spatial and racial/ethnic distribution of experiences of child welfare system involvement in the USA. Second, we detail limitations to this approach that present potential challenges to these estimates. Third, we discuss how these data challenges—while limiting our ability to fully capture child welfare system risks and exposure—provide clear directions toward a research agenda that may elucidate and even mitigate these challenges.

2.1 Estimating Prevalence and Timing of Child Protective Services Involvement

2.1.1 AFCARS and NCANDS

In recent years, researchers have leveraged administrative data, often at the national level, but occasionally at the state level, for three purposes. First, they have used such data to detail the expansive reach and starkly unequal distribution of child welfare system involvement (Kim et al., 2017; Putnam-Hornstein et al., 2011; Yi et al., 2020). Second, they have investigated the correlates of child welfare involvement (Cancian et al., 2013; Edwards, 2019; Putnam-Hornstein & Needell, 2011). Finally, they have considered the potential consequences of child welfare system contact (Berger et al., 2009; Putnam-Hornstein, 2011; Putnam-Hornstein & King, 2014). This contemporary understanding of lifetime risks and prevalence of child welfare involvement and its consequences has been enabled by centralized holdings of administrative records from child protective service and foster care agencies.

At the subnational level, researchers have estimated population-level risks and prevalence of child welfare system contact and examined outcomes of child welfare-involved youth using administrative data from state and local child welfare agencies and jurisdictions (e.g., Edwards et al., 2021a, b; Yi et al., 2023). Additional analyses use linked cross-system administrative data systems, including, for example, the Wisconsin Multi-Sample Person Files (e.g., Font et al., 2018) and the California Children's Data Network (e.g., Putnam-Hornstein et al., 2011) and other birth match linkages within particular states and local jurisdictions (Jonson-Reid & Drake, 2008; Jutte et al., 2011; Shaw et al., 2013). Such data have been especially critical to recent work tying child welfare system contact to economic hardship and insecurity (e.g., Hook et al., 2016), childhood and adult health (e.g., Rubin et al., 2004), and juvenile and adult criminal justice system involvement (e.g., Jonson-Reid & Barth, 2000; Vidal et al., 2017), among many other outcomes of interest.

At the national level, these population-level prevalence estimates have relied on data from the National Child Abuse and Neglect Data System (NCANDS) and the

Adoption and Foster Care Analysis and Reporting System (AFCARS). The NCANDS and AFCARS are furnished by the National Data Archive on Child Abuse and Neglect (NDACAN), housed at Cornell University in Ithaca, NY, in collaboration with Duke University in Durham, NC. The NCANDS and AFCARS are data collection systems that gather information on reports of child maltreatment and foster care, respectively. The reporting and data collection procedures for these two data sources differ, so we discuss each of these data sources in turn.

Reporting to the NCANDS is voluntary and includes case-level data from all 50 states, the District of Columbia, and Puerto Rico on child protective services reports that led to an investigation or alternative response determination. Although early years of the NCANDS data had lower levels of participation, since 2010, nearly all states have submitted data to NCANDS. The NCANDS data includes case-level details for each report of child abuse or neglect that received a CPS response (disposition to an investigation or assessment) in a given reporting year. The NCANDS records include information about child and original caregiver characteristics (e.g., race/ethnicity, sex, disability, parental substance use), details of the maltreatment referral or report (e.g., alleged maltreatment type, alleged maltreatment perpetrator, date of report, county of report), information about how the report was processed (e.g., types of services administered, case disposition), as well as a unique but anonymized state identifier for each child that appears within the reporting state's child protective services system.

The AFCARS includes case-level information on children who are in foster care or have been adopted through agencies that are funded through federal reimbursements accessed through Title IV-E of the Social Security Act. States submit six-month AFCARS files on a semi-annual basis that reflect the most recent case-level activity (i.e., based on the most recent removal) for youth who were in care, or who exited care, during the reporting period. The AFCARS records also include information about child and original caregiver characteristics, details of the case (e.g. reason for foster care, placement and caregiver characteristics, dates of placements and adoption, funding source), as well as a unique and anonymized state identifier for each child appearing within the reporting state's foster care and adoption system. The AFCARS and NCANDS data can be linked through the unique AFCARS identifying variable. Unlike the NCANDS, AFCARS reporting is mandatory for all state and tribal agencies. NCANDS data are collected annually while AFCARS data are collected semi-annually and then collapsed into annual files once submitted to NCANDS.

2.1.2 Synthetic Cohort Life Tables

As annual data, the NCANDS and AFCARS are cross-sectional, meaning that they do not provide information about individual children or families in the child protective services and foster care systems, respectively, *over time*. However, the state identifiers in the NCANDS and AFCARS make it possible to link children observed in a state's records more than once. Further, both the NCANDS and AFCARS include data elements reflecting prior history of child welfare contact within the state—prior substantiated CPS report and prior placement history, respectively—making it possible to determine which children had and had not previously experienced such a child welfare event in the same state. These identifiers, in combination with the synthetic cohort life table method, allow researchers to leverage cross-sectional administrative data to answer research questions about lifetime prevalence and distribution of child welfare system involvement.

In the absence of birth cohort data that allow individual children to be observed from birth until adulthood (the period during which they are at risk of first entering the child welfare system), synthetic cohort or period estimates use single-year population totals and totals of events to estimate risks using hypothetical or synthetic cohorts. Although the synthetic cohort life table method is described extensively elsewhere (e.g., Preston et al., 2001; Sabol et al., 2004; Yi et al., 2020), the construction of and assumptions behind the synthetic cohort life table provide critical insight into the limitations and potential future directions for child welfare research using administrative data. As such, we provide an overview of this method here, in some level of detail.

The synthetic cohort life table is a classic non-parametric demographic and population accounting technique used to estimate prevalence or risk and timing of events in the absence of longitudinal population-level data. This approach has historically been used for analyses of core demographic processes, such as life expectancy at birth and all-cause and cause-specific mortality risks (e.g., Bell & Miller, 2005; Chiang, 1984; Milne, 1815). More recently, synthetic cohort life tables have been used to examine other social processes and experiences, including the risks of experiencing parental prison incarceration (Wildeman, 2009), child poverty (Rank & Hirschl, 1999), and, as discussed here, child welfare system contact (e.g., Yi et al., 2020).

The general synthetic cohort life table relies on two main inputs: (1) counts of observation units first experiencing the event or exposure of interest by time interval and (2) counts of total observation units in each time interval. Taking the specific case of foster care placement as an example, a synthetic cohort life table estimating children's cumulative risk of foster placement would require age-specific (1) numbers of children experiencing a first foster care placement and (2) child population totals for the ages of zero years (birth) through age 18. If examining subgroups of the child population, such as risks for children from different racial/ethnic groups, for example, these counts would need to be specific to those groups.

These two inputs are then converted into age-specific rates through simple division that calculates the proportions of children in each age group that are placed in foster care at that age. Next, these age-specific rates are converted into statistical probabilities of foster care placement (and probabilities of non-placement) at each age. These probabilities differ from the base rate calculations as the probabilities account for the potentially uneven temporal distribution of events, in this case, first foster care placements, over each interval or year of childhood.

At this stage, the probabilities permit the estimation of survival and hazard functions of the likelihood of foster care placement. To do so, probabilities are converted into age-specific risks of foster care placement by applying the probabilities to a hypothetical (or synthetic) cohort that is conventionally set to a starting population, or radix, of 100,000 persons. The probabilities are converted to risks by multiplying the age-specific probabilities by the age-specific population in each age interval of the life table, in sequence. First, the probability of foster care placement at age zero would be multiplied by the hypothetical starting population of 100,000 person to determine the likely number of persons who would experience a foster care placement at age zero. For the next age interval, the population would only include those who hadn't yet experienced foster care placement as per the probability applied at the age zero interval (100,000 less the number of people estimated to be likely to be placed in foster care in the prior step). This population would then be multiplied by the age-specific probability of foster care placement in the next age interval (interval of age 1–2, for example), and so on, and so forth. From an analytic perspective, this starting population size represented by the synthetic cohort is arbitrary and only determines the scaling factor of the life table. For each age interval, the age-specific risk of foster care placement is applied to the radix after accounting for the declining population at risk of experiencing foster care placement for the first time: once a child has been placed in foster care for the first time, they cannot be at risk of experiencing it for the first time at a later age. This calculation is applied to the declining radix for each age interval or row of the life table for the duration of the observation period.

Ultimately, this interval-by-interval procedure calculates both age-specific and cumulative risks of foster care placement and total numbers of children experiencing foster care placement and children who do not (and then age into the population at risk for the next age interval) for the synthetic cohort. Using this synthetic cohort approach, researchers investigating child maltreatment, foster care, and child welfare system contact more generally, have estimated children's overall, race/ethnicity-specific, and state- or county-level risks of experiencing all types of child welfare system events. As presented in Fig. 2.1, the most recent national synthetic cohort or period life table analyses of the AFCARS and NCANDS data estimate that approximately 1 in 3 children will ever experience a maltreatment investigation (Kim et al., 2017), about 1 in 9 children will ever experience confirmed maltreatment (Yi et al., 2020), about 1 in 19 children will ever be placed in foster care (Yi et al., 2020), and about 1 in 100 children will experience termination of parental rights (Wildeman et al., 2020), with substantial racial/ethnic variation in lifetime risks of all of these outcomes.

There are two critical assumptions behind the synthetic cohort life table. First, these estimates assume that age-specific rates of foster care placement are stable over time, or the time it would take for a synthetic cohort member (1 of the 100,000 in the starting radix) to live or age through the life table. Should these age-specific rates of foster care placement be volatile, it would not be plausible for an individual to experience the observed age-specific probabilities over the years examined, or as in this example, over the period of birth to 18 years of age. Although this assumption

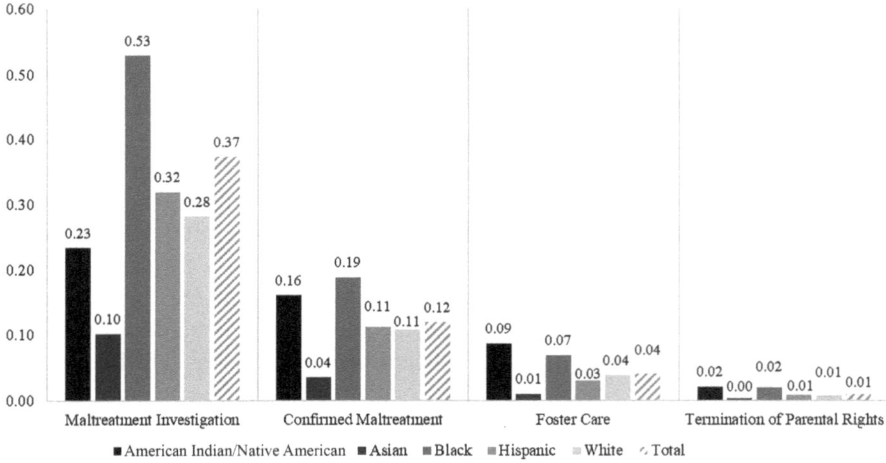

Fig. 2.1 Cumulative risks of child welfare system contact by race/ethnicity. (Sources: Estimates for maltreatment investigation risks are from Kim et al. (2017); for confirmed maltreatment and foster care placement from Yi, Edwards, and Wildeman et al. (2020); and for termination of parental rights from Wildeman et al. (2020))

is flawed during periods of rapid social change, this limitation can be addressed by producing estimates using multiple years of data, for pooled estimates that smooth over year-to-year fluctuations and for year-specific estimates for assessment of volatility or gradual changes in observed rates over the period of interest. Second, the synthetic cohort life table assumes that the first events for all observation units—in the hypothetical example detailed below, any and all children's first foster care placements—are all counted completely (e.g., none are missed) and counted uniquely (e.g., none are double-counted). In other words, the strategy requires certainty that all events of interest are observed for all observation units once, but not multiple times. When applying these assumptions to analyses of AFCARS and NCANDS data, the first assumption, that of stability in age-specific rates of child welfare events, can be assessed, to some degree, by examining observed rates in the administrative data used for the synthetic cohort life table analysis. This approach was discussed regarding the AFCARS data analyzed, for example, by Wildeman and Emanuel (2014), and for data from Cuyahoga County, Ohio, analyzed by Sabol et al. (2004). However, the second of these assumptions is not so easily examined or addressed—particularly when considering child and family mobility, as discussed in the next section.

2.2 Children's Internal Migration

The centralization of the NCANDS and AFCARS data and their analysis via synthetic cohort life tables have fundamentally advanced our understanding of the reach and distribution of child welfare system involvement in the USA. However, the lack of harmonization *across* state-reporting agencies presents a critical challenge to the accuracy of these estimates. More specifically, while the anonymized child identifiers in the AFCARS and NCANDS currently allow researchers to observe individual children over time *within* each state system, they do not uniquely identify child welfare-involved youth *across* state systems.

As children do not necessarily remain in place for the duration of childhood, this may be problematic to existing estimates of prevalence and risk of child welfare system contact. Substantively, this means that our estimates of the prevalence of these experiences—both those estimated using synthetic cohort methods and those that observe birth cohorts within states—may be biased due to child migration within the USA that is not observed in the data. Further, the spatial and social patterning of internal migration (e.g., Bernard et al., 2014; Cooke et al., 2016; Flippen, 2013; Rosenbloom & Sundstrom, 2004) likely affects the direction and magnitude of this bias heterogeneously, with potential implications for estimates of racial/ethnic disparities and disproportionalities in system involvement.

2.2.1 Children's Internal Migration and Bias in Prevalence Estimates

Migration contributes to bias in both birth cohort estimates, which follow individuals over time, and synthetic cohort estimates, which are based on only one year of child welfare data. In order to illustrate how this bias emerges in each case, we will provide one example of a fictional child, Joseph, who has lived in three states. We will then highlight how bias emerges in estimates from birth cohort life tables and synthetic cohort life tables under these conditions. Joseph was born in Massachusetts and lived there until he was three. At that point, he emigrated from Massachusetts and immigrated to New Hampshire, where he lived until he was nine. He then immigrated to New York and stayed there until he turned 18. Joseph has been placed in foster care three times—when he was four and six in New Hampshire, and when he was 11 in New York. He was never placed in foster care in Massachusetts.

Birth cohort estimates of Joseph's foster care experiences would (accurately) indicate that he had never been placed in foster care in Massachusetts. However, because he was not in New Hampshire or New York at the time of his birth—the entry point to the dataset for a birth cohort analysis—he would not be counted as experiencing foster care placement in either of those states. Thus, birth cohort estimates generally tend to present a conservative, or negatively biased, estimate of the cumulative prevalence of foster care placement for children in any given state.

The story is more complicated for synthetic cohort life tables for Joseph, as all 18 years of Joseph's life would be considered in the analysis, regardless of the states in which he was born and eventually came into contact with the foster care system. As with birth cohort life table methods, Joseph would, appropriately, not be counted as having ever been placed in foster care in Massachusetts. He would, however, be appropriately counted as experiencing his first foster care placement in New

Hampshire. Further, in New Hampshire his second placement in that state would *be counted* correctly as a higher-order (i.e., second) placement. This correct identification would be possible because his initial and true first placement in New Hampshire would be known to the child welfare system in New Hampshire, enabled by that state system's access to Joseph's full child welfare history.

His first foster care placement in New York at the age of 11, however, would be problematic. Although this was his first foster care placement in the state of New York, it was not the first foster care placement experienced *in his lifetime*. As such, youth with foster care as well as migration histories can potentially be counted as experiencing first foster care placements multiple times. This is problematic both for state estimates (for New York, in this example) and for national estimates, which would count Joseph as having experienced foster care placement for the first time twice, an impossibility:once in New Hampshire and once in New York. As such, our inability to observe children across states introduces upward bias to national estimates of the cumulative prevalence of child welfare system involvement by multiply counting some children.

But how biased are these existing estimates of lifetime risks of child welfare system contact, and how does this bias vary spatially and socially, across states and racial/ethnic groups?[1] To provide preliminary answers to these questions, we describe patterns of child internal migration in the USA estimated using data from the 2018 American Community Survey (Ruggles et al., 2020). The summary of state-level immigration, emigration, and net migration rates among children presented in Table 2.1 draws attention to two important features of child migration

Table 2.1 Immigration, emigration, and net migration rates per 100,000 children, 2018

	Mean (SD)	Minimum	Median	Maximum
Immigration	1846.93	803.29	1739.71	4453.71
	(764.54)	New York	Texas	North Dakota
Emigration	1846.93	854.56	1680.32	6683.52
	(720.89)	Maine	Alabama	District of Columbia
Net Migration	0.00	−3685.33	118.67	1231.15
	(566.51)	District of Columbia	Ohio	Idaho

Notes: $N = 51$ (50 states and the District of Columbia). Estimates are based on analysis of 2018 data from the American Community Survey (Ruggles et al., 2020). Estimates are weighted by state child population and reported per 100,000 children. States and jurisdictions corresponding to the descriptive statistic reported are presented for the weighted minimum, median, and maximum state migration rates

[1] Wildeman (2018) conducts a highly relevant, though indirect, assessment of cumulative prevalence estimates of child welfare system contact estimated using birth cohort and AFCARS and NDACAN data. The author estimates children's cumulative risks of maltreatment report, investigation, confirmed maltreatment, and foster care placement by age five using birth cohort analysis of data from California and synthetic cohort analysis of California state records in AFCARS and NCANDS data. The author finds slight upward bias (0.4 to 1.9 percentage points) in estimated risks of all three outcomes.

in the USA. First, the estimates show that rates of migration in childhood are substantial.

Table 2.1 also shows that rates vary widely across states, meaning that unobserved child migration is likely to introduce bias in estimates of prevalence at lower levels of jurisdiction (e.g., regional, state), in addition to positively biasing national estimates of system involvement. However, the severity and direction of bias in both group-specific and subnational estimates are not likely to be uniform. Figure 2.2 confirms this, illustrating the wide heterogeneity of net migration rates in the USA across states and racial/ethnic groups (summarized for the total child population in Table 2.1 and detailed in Table 2.2). Children who are non-Hispanic and Black have the widest range in state net migration rates, ranging from −15,648 per 100,000 in Hawai'i to 12,021 per 100,000 in New Mexico. The smallest range in state net migration rates is observed for non-Hispanic White children, ranging from −8360 per 100,000 in Hawai'i to 1609 per 100,000 in Wyoming.

Variation in net migration rates points to the potentially heterogeneous impacts of child migration on the accuracy of state and national risks of child welfare system involvement. However, these rates mask tremendous variation in patterns of immigration and emigration into and out of states that are informative for understanding the ways in which existing estimates could be biased. Child emigration or out-migration rates vary tremendously, as well. In fact, they vary even more than immigration rates. As shown in Table 2.1, for example, the highest state child immigration rate—4454 per 100,000 in North Dakota—was 5.5 times higher than the lowest observed state child immigration rate of 803 per 100,000 in New York. This range is even more dramatic in examining state rates of child emigration. The District of Columbia was the jurisdiction with the highest rate of children leaving the state, with 6684 children emigrating per 100,000, a rate that was nearly eight times as high as the child emigration rate in Maine (855 per 100,000; Table 2.1).

The combinations of immigration and emigration flows that result in the net migration rates observed in Fig. 2.2 vary tremendously, as well. For example, while two states may have similar net migration rates—for example Iowa (505 per 100,000) and Texas (524 per 100,000)—the volume of children moving into and out of those states indicates that the population dynamics resulting in that net change in the child population are quite different. In Iowa, the child immigration rate in 2018 was 2212 per 100,000 and the child emigration rate was 1708 per 100,000. In Texas, those rates were 1740 per 100,000 and 1216 per 100,000. Further, just as net migration rates vary socially as well as spatially (Fig. 2.2), immigration rates and emigration rates do as well.

Figure 2.3 presents immigration and emigration rates per 100,000 children for all children and by racial/ethnic group for a selection of states that illustrate the wide range in these dynamics across space and society; the full set of estimates is presented in Tables 2.3 and 2.4. Although it was impossible to visually present immigration and emigration rates for all racial/ethnic groups in all states simultaneously, this subset illustrates the intersection of state and racial/ethnic heterogeneity that is likely to shape the direction and magnitude of bias in state and group-specific life

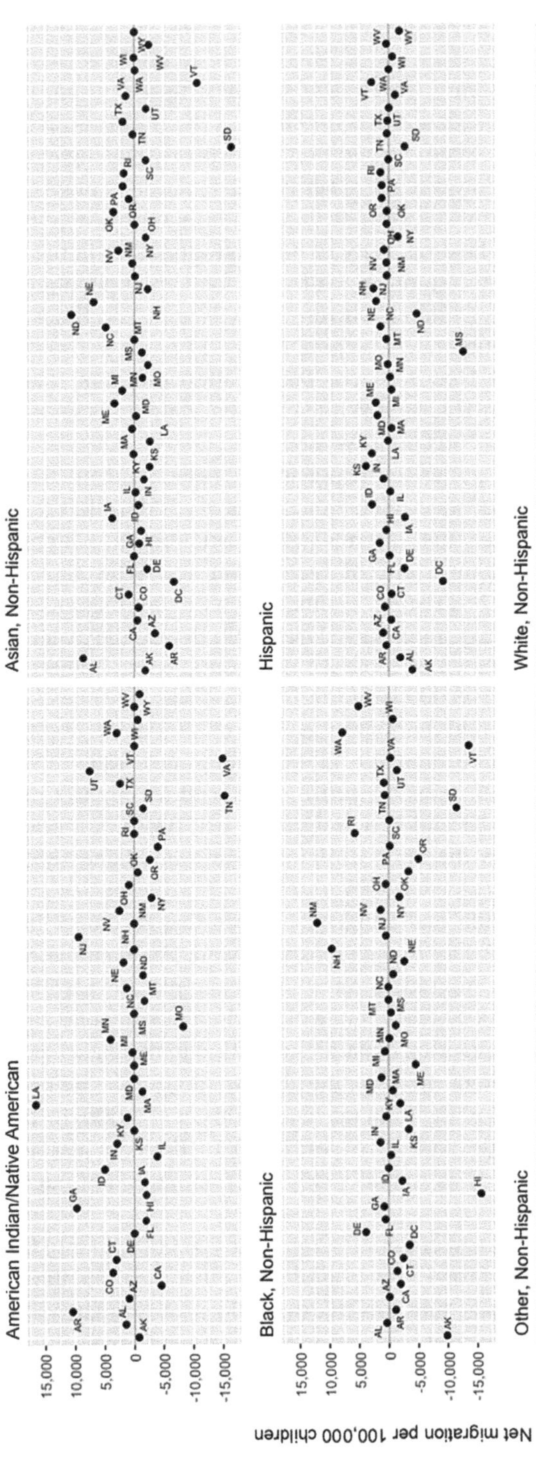

Fig. 2.2 Net migration rates per 100,000 children by race/ethnicity and state. (Notes: Analysis of 2018 data from the American Community Survey (Ruggles et al., 2020))

Table 2.2 Net migration rates per 100,000 children by race/ethnicity and state

State	Overall	American Indian/ Native American	Asian	Black	Hispanic	White
Alabama	590	1453	8597	377	−1933	1005
Alaska	−1292	−720	−1891	−9852	−3975	−2634
Arizona	817	936	−3555	−78	962	1351
Arkansas	−105	10,430	−5926	−1139	423	218
California	−624	−4585	−484	−1956	−397	−789
Colorado	69	3660	−701	−1371	678	22
Connecticut	−556	3086	970	−2425	−465	−601
Delaware	−288	0	−2195	3866	−2634	−1556
District of Columbia	−3685	na	−6721	−3461	−9190	1057
Florida	277	−1990	58	574	−122	267
Georgia	916	9696	−818	783	1549	645
Hawai'i	−1965	−2071	−1136	−15,648	410	−8360
Idaho	1231	4968	−677	0	2830	859
Illinois	−459	−3960	−209	−374	−252	−583
Indiana	−60	2898	−1662	1380	901	−378
Iowa	505	16,521	−2689	−1954	142	338
Kansas	−324	−1811	3732	−2269	−2722	112
Kentucky	302	0	−2635	−3401	3867	426
Louisiana	11	1146	141	399	2860	−681
Maine	705	0	3355	−4567	2203	652
Maryland	268	0	−289	1185	1922	−634
Massachusetts	−425	−1424	383	−711	−449	−271
Michigan	189	258	2045	618	−447	133
Minnesota	278	3963	−1410	−128	−199	461
Mississippi	−992	0	−1282	−419	−12,576	−241
Missouri	11	−8344	−2337	−1206	179	538
Montana	905	−1791	0	0	425	1029
Nebraska	−481	1780	6844	−2651	2161	−1439
Nevada	1165	2483	2684	1339	832	1503
New Hampshire	410	0	−2305	9574	2599	77
New Jersey	88	9326	−98	448	346	−141
New Mexico	−137	−10	353	12,021	403	−1445
New York	−1077	−3000	−1942	−1808	−1516	−439
North Carolina	341	1216	4817	25	1403	−269
North Dakota	−1018	−1538	10,622	−783	−4678	−1300
Ohio	119	899	−30	495	372	25
Oklahoma	−55	−610	3533	−3340	345	−51
Oregon	−67	−2728	935	−5015	1177	−378
Pennsylvania	140	−4048	1934	−176	1173	−186
Rhode Island	322	0	1779	5724	1401	−754
South Carolina	231	0	−1954	−102	78	379

(continued)

Table 2.2 (continued)

State	Overall	American Indian/ Native American	Asian	Black	Hispanic	White
South Dakota	−1439	−1518	−16,383	−11,479	−2672	−814
Tennessee	398	−15,298	263	686	313	477
Texas	524	2394	1947	821	202	726
Utah	−138	7489	−1998	−1360	−43	−284
Vermont	358	0	−10,644	−13,514	2888	480
Virginia	−299	−14,933	1449	−273	−1085	−382
Washington	222	2981	−100	7841	6	21
West Virginia	−77	0	−2503	5134	374	−196
Wisconsin	216	−551	76	−666	−624	486
Wyoming	218	−887	0	na	−1799	1609

Notes: Analysis of 2018 data from the American Community Survey (Ruggles et al., 2020)

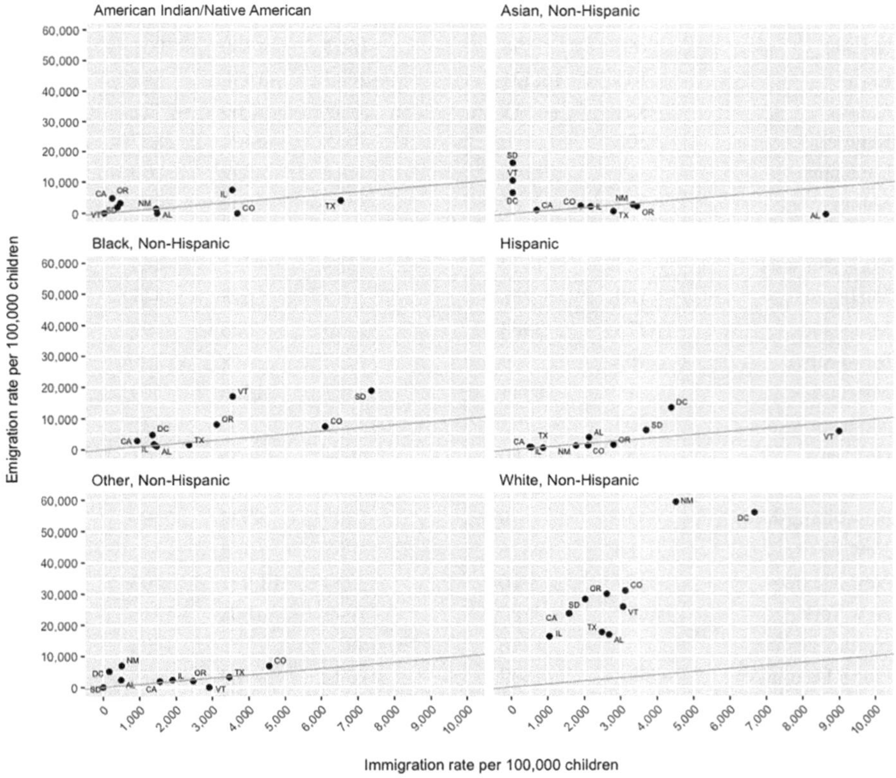

Fig. 2.3 Immigration and emigration rates per 100,000 children by race/ethnicity and state, select states. (Notes: Analysis of 2018 data from the American Community Survey (Ruggles et al., 2020))

Table 2.3 Immigration rates per 100,000 children by race/ethnicity and state

State	Overall	American Indian/ Native American	Asian	Black	Hispanic	White
Alabama	2271	1453	8597	1457	2113	2687
Alaska	3896	308	0	0	2828	4548
Arizona	3004	1489	2748	4062	1995	4225
Arkansas	2156	11,115	681	1665	2727	2068
California	870	215	662	924	496	1580
Colorado	2967	3660	1864	6093	2087	3127
Connecticut	2014	3086	5623	2594	1871	1509
Delaware	2139	0	2254	5430	529	1056
District of Columbia	2998	na	0	1341	4372	6667
Florida	2375	634	2786	1652	1665	3028
Georgia	2246	10,251	1779	1957	2681	2141
Hawai'i	4170	393	548	14,566	6929	12,530
Idaho	3643	5298	1834	0	4261	3529
Illinois	1070	3518	2135	1385	541	1048
Indiana	1656	2898	4300	2704	2892	1107
Iowa	2212	17,268	0	3735	1319	1798
Kansas	2830	453	7308	6196	2450	2354
Kentucky	2274	0	451	3239	5455	1963
Louisiana	1625	1571	1811	1455	4143	1297
Maine	1560	0	3355	0	7110	1366
Maryland	2353	0	2474	2046	4752	1769
Massachusetts	1050	0	3286	678	1210	887
Michigan	1095	258	4017	1483	1386	875
Minnesota	1196	5300	149	1352	1469	1085
Mississippi	1890	0	1952	1663	919	2186
Missouri	1962	0	3035	1860	2696	1956
Montana	3080	0	0	0	4353	3316
Nebraska	2177	1780	12,901	7033	3160	1230
Nevada	3344	3065	4519	3437	2467	4071
New Hampshire	2022	0	0	9574	4706	1524
New Jersey	1553	14,693	2006	2258	1586	1188
New Mexico	2478	1426	3313	12,021	1755	4503
New York	803	376	695	1137	705	762
North Carolina	2553	1486	6711	1837	2937	2341
North Dakota	4454	5827	11,776	1634	12,109	3940
Ohio	1250	899	4093	1556	1812	1022
Oklahoma	2225	176	6376	1448	2008	2643
Oregon	2656	437	3420	3112	2795	2620
Pennsylvania	1439	0	4917	1039	2643	980
Rhode Island	2698	0	2587	6768	2790	2402
South Carolina	2908	0	231	2068	1951	3459

(continued)

Table 2.3 (continued)

State	Overall	American Indian/ Native American	Asian	Black	Hispanic	White
South Dakota	1923	360	0	7354	3691	2014
Tennessee	2560	0	4540	2199	2699	2610
Texas	1740	6495	2775	2354	864	2489
Utah	2551	7857	1126	3320	2209	2575
Vermont	3176	0	0	3547	8977	3067
Virginia	2829	2073	3014	2390	2421	2844
Washington	2993	3007	1962	8716	2810	2856
West Virginia	2054	0	0	6167	7113	1758
Wisconsin	1412	551	2825	994	1151	1430
Wyoming	3833	1226	0	na	3630	4072

Notes: Analysis of 2018 data from the American Community Survey (Ruggles et al., 2020)

table estimates of child welfare contact prevalence. Take, for instance, the state of Alabama. For children in Alabama who are American Indian/Native American, the immigration rate is comparatively moderate (1453 per 100,000) while the emigration rate is zero. For children in the same state who are non-Hispanic and White, however, the immigration and emigration rates are higher, with the child emigration rate being substantially higher (1681 and 2687 per 100,000, respectively). These differences in migration flows show that child migration is likely to have differential impacts on bias in cumulative risk estimates for child welfare system contact across racial/ethnic groups, even within the same state.

2.3 Potential Solutions and Future Directions

Future efforts to address these biases will need to focus not only on intrastate cross-system linkages but also on inter-state linkages. As illustrated in this discussion, without the ability to observe children across state lines and jurisdictions, estimates of prevalence and risks of child welfare system contact will continue to be affected by potential bias due to current data's inability to account for internal migration in the USA. However, these limitations point directly to next steps and research objectives for continued investigation and expanded data resources for child welfare research.

The most obvious solution would be to construct a fully linked administrative dataset of all children in the USA. This would allow unique children to be identifiable in different and multiple state child welfare systems across the country, eliminating the bias issues in estimating prevalence and distribution of child welfare system contact described earlier. The construction of such a database would require collaboration and coordination across child welfare agencies and allow comparison and linkage of identifying information for children across state systems. Absent the availability of a fully linked administrative dataset of all children in the USA and

Table 2.4 Emigration rates per 100,000 children by race/ethnicity and state

State	Overall	American Indian/ Native American	Asian	Black	Hispanic	White
Alabama	1680	0	0	1080	4046	1681
Alaska	5188	1027	1891	9852	6803	7182
Arizona	2187	553	6303	4139	1033	2874
Arkansas	2261	684	6607	2805	2304	1850
California	1494	4799	1146	2880	894	2369
Colorado	2897	0	2565	7464	1409	3104
Connecticut	2570	0	4653	5019	2336	2110
Delaware	2427	0	4448	1564	3163	2612
District of Columbia	6684	na	6721	4802	13,562	5610
Florida	2098	2624	2729	1078	1787	2761
Georgia	1330	555	2597	1174	1132	1495
Hawai'i	6135	2464	1684	30,214	6519	20,890
Idaho	2411	329	2510	0	1432	2670
Illinois	1529	7478	2344	1758	792	1631
Indiana	1716	0	5963	1324	1992	1485
Iowa	1708	747	2689	5689	1177	1459
Kansas	3154	2264	3576	8465	5171	2243
Kentucky	1972	0	3086	6640	1588	1537
Louisiana	1615	425	1670	1056	1284	1978
Maine	855	0	0	4567	4907	714
Maryland	2085	0	2763	861	2830	2403
Massachusetts	1475	1424	2903	1389	1659	1158
Michigan	906	0	1972	865	1833	742
Minnesota	918	1338	1559	1480	1668	624
Mississippi	2882	0	3234	2083	13,495	2426
Missouri	1950	8344	5372	3066	2517	1418
Montana	2176	1791	0	0	3928	2287
Nebraska	2657	0	6057	9684	999	2669
Nevada	2179	583	1834	2099	1635	2568
New Hampshire	1612	0	2305	0	2107	1446
New Jersey	1465	5367	2104	1810	1240	1330
New Mexico	2614	1436	2960	0	1353	5949
New York	1880	3376	2636	2944	2221	1200
North Carolina	2211	269	1894	1812	1534	2610
North Dakota	5471	7366	1154	2417	16,787	5240
Ohio	1131	0	4123	1061	1440	998
Oklahoma	2280	786	2843	4788	1663	2693
Oregon	2723	3165	2485	8128	1619	2997
Pennsylvania	1298	4048	2983	1216	1471	1166
Rhode Island	2376	0	807	1045	1389	3156
South Carolina	2677	0	2185	2170	1874	3081

(continued)

Table 2.4 (continued)

State	Overall	American Indian/ Native American	Asian	Black	Hispanic	White
South Dakota	3362	1879	16,383	18,834	6363	2828
Tennessee	2162	15,298	4277	1513	2386	2133
Texas	1216	4100	828	1533	662	1763
Utah	2689	368	3123	4680	2252	2859
Vermont	2818	0	10,644	17,061	6089	2587
Virginia	3127	17,006	1565	2662	3506	3226
Washington	2771	26	2061	875	2804	2835
West Virginia	2131	0	2503	1033	6739	1953
Wisconsin	1196	1103	2749	1660	1774	944
Wyoming	3615	2113	0	na	5429	2463

Notes: Analysis of 2018 data from the American Community Survey (Ruggles et al., 2020)

their residential and child welfare histories, analyses of existing data have much to offer in taking stock of and even accounting for children's internal migration in the USA and its implications for estimates of the prevalence of child welfare system involvement.

First, existing data, including such tools as the American Community Survey, analyzed above, can be used to more comprehensively document patterns of child migration in the United States. Although the estimated migration rates presented in this chapter present a mechanical and conceptual introduction to the introduction of bias into the best available estimates of the prevalence of child welfare system involvement, a more detailed examination is needed. As shown in the child migration estimates above (see Table 2.1, Figs. 2.2 and 2.3), child migration rates vary widely across states and across racial/ethnic groups. However, due to the age structure and gendered, racialized, and socioeconomically unequal composition of the child welfare-involved population, an understanding of the bias of migration on estimates of risks of involvement will require a consideration of how migration is distributed across other dimensions of social difference and disproportionality, as well. For example, very young children who are less than one year old have the highest child maltreatment substantiation rate of all age groups (25 per 1000 children; Child Welfare Information Gateway 2019). For lifetime risks of different types of child welfare system contact, then, the estimates might be more heavily biased by internal migration of young children. A more detailed analysis of the distribution and demographics of child migration will therefore be critical to addressing the challenge of migration bias in prevalence estimates.

Finally, using this more comprehensive understanding of child internal migration, existing county-level data from the AFCARS and NCANDS can be used to examine impacts of migration on bias in county-level estimates of child welfare system involvement. The task at hand would be to produce "naïve" or "blind" estimates that treat children's between-county migration histories as unknown and compare them to fully informed estimates that use the aforementioned unique state child identifiers to detect children who appear in the child welfare system in multiple counties in the same state. Discrepancies between the naïve and fully informed county-level estimates could give research a sense of the magnitude of bias driven by child migration at the local level and ultimately inform the development of methodological bias corrections for life table estimation of lifetime risks of child welfare system contact at the state and national levels.

Acknowledgments The authors would like to recognize and thank Ashley Lewis and Sarah Sernaker for their excellent research assistance.

References

Andersen, S. H., & Wildeman, C. (2014). The effect of paternal incarceration on children's risk of foster care placement. *Social Forces, 93*(1), 269–298. https://doi.org/10.1093/sf/sou027

Bell, F. C., & Miller, M. L. (2005). *Life tables for the United States Social Security Area 1900–2100.* Social Security Administration. Available at: https://www.ssa.gov/OACT/NOTES/pdf_studies/study120.pdf

Berger, L. M., Bruch, S. K., James, S., Johnson, E. I., & Rubin, D. (2009). Estimating the 'impact' of out-of-home placement on child well-being: Approaching the problem of selection bias. *Child Development, 80*(6), 1856–1876. https://doi.org/10.1111/j.1467-8624.2009.01372.x

Bernard, A., Bell, M., & Charles-Edwards, E. (2014). Life-course transitions and the age profile of internal migration. *Population and Development Review, 40*(2), 213–239. https://doi.org/10.1111/j.1728-4457.2014.00671.x

Cancian, M., Yang, M., & Slack, K. S. (2013). The effect of additional child support income on the risk of child maltreatment. *Social Service Review, 87*(3), 417–437. https://doi.org/10.1086/671929

Chiang, C. L. (1984). *The life table and its applications*. Krieger.

Cooke, T. J., Mulder, C. H., & Thomas, M. (2016). Union dissolution and migration. *Demographic Research, 34*, 741–760. https://doi.org/10.4054/DemRes.2016.34.26

Edwards, F. (2019). Family surveillance: Police and the reporting of child abuse and neglect. *RSF: The Russell Sage Foundation Journal of the Social Sciences, 5*(1), 50–70. https://doi.org/10.7758/RSF.2019.5.1.03

Edwards, F., Wakefield, S., Healey, H., & Wildeman, C. (2021a) Contact with Child Protective Services is pervasive but unequally distributed by race and ethnicity in large US counties. *Proceedings of the National Academy of Sciences 118*(30), e2106272118. https://doi.org/10.1073/pnas.2106272118

Edwards, F., Wakefield, S., Healy, K., & Wildeman, C. (2021b). Correction for Edwards et al. Contact with Child Protective Services is pervasive but unequally distributed by race and ethnicity in large US counties. *Proceedings of the National Academy of Sciences 118*(42), e2106272118. https://doi.org/10.1073/pnas.2116639118

Fallesen, P. (2014). Identifying divergent foster care careers for Danish children. *Child Abuse & Neglect, 38*(11), 1860–1871. https://doi.org/10.1016/j.chiabu.2014.08.004

Flippen, C. (2013). Relative deprivation and internal migration in the United States: A comparison of Black and White men. *American Journal of Sociology, 118*(5), 1161–1198. https://doi.org/10.1086/668691

Font, S. A., Berger, L. M., Cancian, M., & Noyes, J. L. (2018). Permanency and the educational and economic attainment of former foster children in early adulthood. *American Sociological Review, 83*(4), 716–743. https://doi.org/10.1177/0003122418781791

Hook, J. L., Romich, J. L., Lee, J. S., Marcenko, M. O., & Kang, J. Y. (2016). Trajectories of economic disconnection among families in the child welfare system. *Social Problems, 63*, 161–179. https://doi.org/10.1093/socpro/spw006

Jonson-Reid, M., & Barth, R. P. (2000). From maltreatment report to juvenile incarceration: The role of child welfare services. *Child Abuse & Neglect, 24*(4), 505–520. https://doi.org/10.1016/s0145-2134(00)00107-1

Jonson-Reid, M., & Drake, B. (2008). Multisector longitudinal administrative databases: An indispensable tool for evidence-based policy for maltreated children and their families. *Child Maltreatment, 13*(4), 392–399. https://doi.org/10.1177/1077559508320058

Jutte, D. P., Roos, L. L., & Brownell, M. D. (2011). Administrative record linkage as a tool for public health research. *Annual Review of Public Health, 32*, 91–108. https://doi.org/10.1146/annurev-publhealth-031210-100700

Kim, H., Wildeman, C., Jonson-Reid, M., & Drake, B. (2017). Lifetime prevalence of investigating child maltreatment among US children. *American Journal of Public Health, 107*, 274–280. https://doi.org/10.2105/AJPH.2016.303545

Milne, J. (1815). *Treatise on the valuation of annuities and assurances on lives and survivorships*. Longman, Hurst, Rees, Orme, and Brown.

Preston, S. H., Heuveline, P., & Guillot, M. (2001). *Demography: Measuring and modeling population processes*. Blackwell.

Putnam-Hornstein, E. (2011). Report of maltreatment as a risk factor for injury death: A prospective birth cohort study. *Child Maltreatment, 16*(3), 163–174. https://doi.org/10.1177/1077559511411179

Putnam-Hornstein, E., & King, B. (2014). Cumulative teen birth rates among girls in foster care at age 17: An analysis of linked birth and child protection records from California. *Child Abuse & Neglect, 38*(4), 698–705. https://doi.org/10.1016/j.chiabu.2013.10.021

Putnam-Hornstein, E., & Needell, B. (2011). Predictors of child protective service contact between birth and age five: An examination of California's 2002 birth cohort. *Children and Youth Services Review, 33*, 1337–1344. https://doi.org/10.1016/j.childyouth.2011.04.006

Putnam-Hornstein, E., Webster, D., Needell, B., & Magruder, J. (2011). A public health approach to child maltreatment surveillance: Evidence from a data linkage project in the United States. *Child Abuse Review, 20*, 256–273. https://doi.org/10.1002/car.1191

Rank, M. R., & Hirschl, T. A. (1999). The economic risk of childhood in America: Estimating the probability of poverty across the formative years. *Journal of Marriage and Family, 61*(4), 1058–1067. https://doi.org/10.2307/354024

Rosenbloom, J. L., & Sundstrom, W. A. (2004). The decline and rise of interstate migration in the United States: Evidence from the IPUMS, 1850-1990. *Research in Economic History, 22*, 289–325. https://doi.org/10.3386/w9857

Rubin, D. M., Alessandrini, E. A., Feudtner, C., Mandell, D. S., Localio, A. R., & Hadley, T. (2004). Placement stability and mental health costs for children in foster care. *Pediatrics, 113*(5), 1336–1341. https://doi.org/10.1542/peds.113.5.1336

Ruggles, S., Flood, S., Goeken, R., Grover, J., Meyer, E., Pacas, J., & Sobek, M. (2020). *IPUMS USA: Version 10.0 [dataset]*. IPUMS.

Sabol, W., Coulton, C., & Polousky, E. (2004). Measuring child maltreatment risk in communities: A life table approach. *Child Abuse & Neglect, 28*, 967–983. https://doi.org/10.1016/j.chiabu.2004.03.014

Shaw, T. V., Barth, R. P., Mattingly, J., Ayer, D., & Berry, S. (2013). Child welfare birth match: Timely use of child welfare administrative data to protect newborns. *Journal of Public Child Welfare, 7*, 217–237. https://doi.org/10.1080/15548732.2013.766822

Staer, T., & Bjørknes, R. (2015). Ethnic disproportionality in the child welfare system: A Norwegian national cohort study. *Children and Youth Services Review, 56*(C), 26–32. https://doi.org/10.1016/j.childyouth.2015.06.008

Vidal, S., Prince, D., Connell, C. M., Caron, C. M., Kaufman, J. S., & Tebes, J. K. (2017). Maltreatment, family environment, and social risk factors: Determinants of the child welfare to juvenile justice transition among maltreated children and adolescents. *Child Abuse & Neglect, 63*, 7–18. https://doi.org/10.1016/j.chiabu.2016.11.013

Vinnerljung, B., Öman, M., & Gunnarson, T. (2005). Educational attainments of former child welfare clients – a Swedish national cohort study. *International Journal of Social Welfare, 14*(4), 265–276. https://doi.org/10.1111/j.1369-6866.2005.00369.x

Wildeman, C. (2009). Parental imprisonment, the prison boom, and the concentration of childhood disadvantage. *Demography, 46*(2), 265–280. https://doi.org/10.1353/dem.0.0052

Wildeman, C. (2018). The incredibly credible prevalence of child protective services contact in New Zealand and the United States. *American Journal of Public Health, 108*(4), 438–439. https://doi.org/10.2105/AJPH.2018.304313

Wildeman, C., & Emanuel, N. (2014). Cumulative risks of foster care placement by age 18 for U.S. children, 2000-2011. *PLoS ONE, 9*(3), e92785. https://doi.org/10.1371/journal.pone.0092785

Wildeman, C., Edwards, F. R., & Wakefield, S. (2020). The cumulative prevalence of termination of parental rights for U.S. children, 2000-2016. *Child Maltreatment, 25*(1), 32–42. https://doi.org/10.1177/1077559519848499

Yi, Y., Edwards, F. R., & Wildeman, C. (2020). Cumulative prevalence of confirmed maltreatment and foster care placement for US children by race/ethnicity, 2011-2016. *American Journal of Public Health*, e1–e6. 2019. https://doi.org/10.2105/AJPH.2019.305554

Yi, Y., Edwards, F., & Wildeman, C. (2023). State-level variation in the cumulative prevalence of child welfare system contact, 2015–2019. *Children and Youth Services Review 147*(106832): https://doi.org/10.1016/j.childyouth.2023.106832

Chapter 3
The Use of Birth Records to Study Child Abuse and Neglect

Emily Putnam-Hornstein, Stephanie Cuccaro-Alamin, and Rhema Vaithianathan

As the chapters in this volume indicate, the use of administrative data to inform child protective services (CPS) has grown exponentially during the last few decades. Recent technological innovations have allowed researchers and agencies to link administrative datasets from systems that serve children and families to answer important questions about CPS policy and practice. Increasingly, the child protection field is focused on how these linked data can be leveraged to identify children and families at greatest risk of child maltreatment in order to focus the system's response on those most in need. This chapter examines how the linkage of birth records to CPS records offers opportunities to advance this work. Birth records are a largely untapped resource for child maltreatment research. Researchers at the Children's Data Network (CDN) have been linking birth records with data from CPS and other systems to better understand child maltreatment risk and outcomes at the population level. This chapter reviews this research, highlighting the benefits of working with vital birth records. We also detail eight important findings learned from utilizing birth records as an anchor for data linkages in child maltreatment research.

E. Putnam-Hornstein (✉)
University of North Carolina at Chapel Hill, Chapel Hill, NC, USA
e-mail: eputnamhornstein@unc.edu

S. Cuccaro-Alamin
California Child Welfare Indicators Project, University of California at Berkeley, Berkeley, CA, USA

R. Vaithianathan
Centre for Social Data Analytics, Auckland University of Technology, Auckland, New Zealand

C. M. Connell, D. M. Crowley (eds.), *Strengthening Child Safety and Well-Being Through Integrated Data Solutions*, Child Maltreatment Solutions Network, https://doi.org/10.1007/978-3-031-36608-6_3

3.1 Misplaced Resources

In December 2014, the National Institutes of Health abruptly announced the cancellation of the National Children's Study. The study, mandated by the Children's Health Act of 2000, was designed to be the gold standard of survey research and would follow 100,000 children prenatally through age 21 to examine the impact of a wide range of factors on child development and health. Researchers planned to study the influence of everything from pollution to psychosocial factors on children's outcomes. The nationally representative sample was designed to be large enough to enable researchers to examine rarer diseases, such as autism and type 1 diabetes. In the end, despite years of meetings, a few pilot sites, and $1.3 billion spent, the study never launched.

Prospective longitudinal surveys, such as that envisioned through the National Children's Study, have long provided a critical method for answering important questions regarding child development. Done well, this study would have produced incredible data that could have led to important discoveries. Although the survey's failure was certainly an example of gross mismanagement, it also reflects how the environment for survey research has changed in recent decades. Nationally representative longitudinal surveys have become increasingly expensive to conduct and difficult to administer. They are often plagued by high nonresponse rates and the findings dated by the time the studies have concluded (Meyer et al., 2015).

Despite these limitations, we continue to spend an exceptional amount of money to collect data inefficiently using surveys. For a moment, imagine a national child study not limited to 100,000 children but involving millions of children who could be longitudinally followed over time from birth. We don't have to imagine because the possibility exists using linked administrative data.

3.2 Linked Administrative Data

Given the limitations of surveys, researchers and policy makers are increasingly utilizing administrative data from systems that serve children and families to augment and, in many cases, replace survey data (Harron, 2016; O'Hara et al., 2017). Administrative data can be used to populate missing or incomplete survey fields, validate existing information, and extend outcome measurement beyond initial data collection (Card et al., 2010; Meyer et al., 2015; Simon, 2014). Because they have already been gathered, administrative data are considerably less expensive than research involving fieldwork or complex sampling techniques, and nonresponse or participant attrition at follow-up is typically not an issue (Christensen, 1958; Dunn, 1946; Meyer et al., 2015; Roos et al., 2008).

Technological innovations such as advances in data storage and record-matching techniques have facilitated the ability of researchers and public agencies to link these large administrative datasets to study child development (Culhane & Metraux,

1997; Enamorado et al., 2019; Hotz et al., 1998; Victor & Mera, 2001). Specifically, by linking event-based encounter data, researchers can construct longitudinal records of children's service trajectories (Connelly et al., 2016; Culhane & Metraux, 1997; Harron, 2016).

3.3 Birth Record Linkages

> Each person in the world creates a Book of Life. This Book starts with birth and ends with death. Its pages are made up of the records of the principal events in life. Record linkage is the name given to the process of assembling the pages of this Book. (Dunn, 1946)

We like to use the analogy of the Book of Life to illustrate this approach. Any individual's story is bound by two universal events: birth and death. Vital birth records provide a powerful anchor for establishing population-based record linkages. As indicated, researchers at the CDN have been establishing such linkages between vital birth records and other service systems data in California to study child maltreatment risk. This chapter describes the birth cohort approach used by the CDN's research team, illustrating the logistic and informational benefits of working with birth records and highlighting eight important findings that have emerged from findings to date.

3.4 The Virtues of Birth Records

Vital birth records are one of the most underutilized sources of administrative data. This is unfortunate, because birth records include a great deal more information than is included in the basic birth certificate form many parents and caregivers bring home from the hospital. A tremendous amount of information is collected at birth; what is found on the birth certificate represents only a small snapshot of that data. Further, birth records are universally collected and nationally standardized; they contain valuable health, demographic, financial, and service information on multiple individuals, and they can serve as a population base, or spine (i.e., a unified, population-level registry), for developing prospective studies regarding children.

3.4.1 Universally Collected (Good for Research and Real-World Applications)

In the Unites States, the reporting of vital events such as birth and death is legally mandated (Schwartz, 2009). As a result, vital records represent the entire universe of such events (all births) rather than a sample. Given this universality, sampling error—a common problem in survey research—is not an issue. Thus, findings associated with birth records have high levels of external validity and provide valuable information for policy making or other real-world applications (Harron, 2016; Harron et al., 2017). Vital records frequently serve as a basis for public health efforts, providing contextualized understanding of current or changing demographic trends in US health or health disparities (National Research Council, 2009). Recent years have seen efforts to better open, integrate, and protect vital records for research purposes. For instance, the Commission on Evidence-Based Policymaking explicitly pointed to vital records, including birth records, in the implementation of the Foundations for Evidence-Based Policymaking Act (Abraham et al., 2017).

3.4.2 Nationally Standardized and Well-Documented Fields

In most states and territories, after a birth occurs, birth records are processed by county public health officials and sent to state databanks. Although maintenance of these public records is the legal responsibility of state agencies, states must report data annually to the National Center for Health Statistics for the Vital Statistics Cooperative Program (Schwartz, 2009). The center sets national standards for consistency, quality, and timeliness of data (Schwartz, 2009). Although not error-free, this national standardization ensures data are uniform across geographies and clients and makes birth records highly reliable (Christensen, 1958).

3.4.3 Information for Three Individuals (Child, Mother, Father)

From a research perspective, birth records provide an efficient and cost-effective way to access valuable information. A single birth record includes information on as many as three individuals: the child and two parents. This record structure enables researchers to construct a family at birth and link administrative data for each member, to follow them through service systems over time. Although the rate of missing paternity records varies across states and localities, most records in California (92%) have two named parents.

3.4.4 Health, Demographic, Financial, and Service Information

Birth records contain a significant amount of health, demographic, financial, and service information. The National Center for Health Statistics (2019) mandates the minimum fields that all states must collect. These rich fields can be used as covariates in analyses of child maltreatment risk or other outcomes. Fields include a child's gender, race and ethnicity, birth weight, Apgar scores (i.e., ranging from 0 to 10 regarding an infant's appearance, heart rate, muscle tone, and other indicators of medical condition at birth), birth complications, medical conditions, and congenital abnormalities. Additionally, information regarding birth location, method of payment (insurance), and characteristics of labor and delivery are included. Parental demographics, including birthplace, education, age, race and ethnicity, prenatal care history, and information on risk factors (e.g., smoking), also are available.

3.4.5 Provides a Population Base (Spine) for Developing Prospective Studies

As the first administrative record for an individual, birth records provide a useful population-based spine for developing prospective studies regarding child maltreatment. A birth cohort spine establishes a foundational population to which records existing in other service and case management systems, and covering different time periods, can be linked. Too often as researchers, our studies are retrospective; we default to looking back instead of employing methods that follow individuals forward in order to study the relationship of risk to outcomes. The birth cohort approach provides this opportunity.

3.5 Child Maltreatment Research: Innovations Using Birth Records

At the CDN, birth records serve as the spine for all administrative data linkage work. CDN researchers have systematically linked the birth records of more than 10 million children born in California during the last decade to CPS records in order to examine health and safety outcomes. The data have also been linked to other administrative sources to examine participation in state benefit programs, including Medi-Cal (the state's Medicaid program), Women, Infants & Children (WIC), Developmental Services, CalWORKs (the state's Temporary Assistance for Needy Families (TANF) program), and CalFresh (the state's Supplemental Nutrition Assistance Program (SNAP)), as well as statewide public education.

The following sections illustrate the benefits of this birth cohort approach, detailing eight important findings learned from utilizing birth records as an anchor for data linkages in child maltreatment research. Although specific to California, these findings highlight the advantages of this approach in the hope it can be replicated in other states.

3.5.1 Calculating Cumulative Rates of Child Protective Services Involvement

One of the most important innovations using the birth cohort approach has been the ability to follow children over time and calculate cumulative rates of CPS involvement. These studies have revealed how common contact with CPS is for children in general. National estimates developed by Kim et al. (2017) using National Child Abuse and Neglect Data System Child Files (2003–2014) and census data showed that over their lifetime, more than 1 in 3 children (37.4%) will be the subject of a reported and investigated child abuse or neglect allegation. The highest lifetime rate was reported for Black children (53.0%) and the lowest rate for Asians and Pacific Islanders (10.2%).

In California, we linked the birth records of all children born between 2006 and 2007 to statewide CPS records to establish the cumulative rate of child welfare system contact by age 5. In total, 14.8% of children had been reported as alleged victims of abuse and neglect by age 5 (Putnam-Hornstein et al., 2021). Extending outcome measurement to age 18, we found that the rate reached approximately 27%. Although lower than other national estimates, these findings are consistent with the direction of previous research and critical to policy discussions. The birth cohort approach may in fact produce a more conservative, lower-bound estimate due to out-migration (i.e., children in our data cannot be administratively "followed" across state lines).

Similar to national synthetic estimates produced by Kim et al. (2017), the cumulative contact rates for our state birth cohorts also revealed pronounced racial and ethnic disparities. For example, in a similar analysis, we found that almost 1 in 2 Black children were investigated for abuse and neglect before age 18 (Putnam-Hornstein et al., 2021). These studies show the power of using birth cohorts to establish cumulative rates of child welfare involvement for entire populations, as well as for highlighting disparities among subpopulations.

3.5.2 Developing Population-Based, Prospective Studies to Study Rare Events

Birth records can be used as the initial spine to develop population-based prospective studies, which are particularly important when exploring rare events such as child maltreatment fatalities. For example, we linked vital birth, child welfare, and death records to examine injury death among all children born in California between 1999 and 2006 (Putnam-Hornstein, 2011). We found that a report to the child welfare system, regardless of whether it was substantiated, was the strongest predictor of an injury death by age 5. After adjusting for risk factors at birth, results showed that children who had a prior allegation of maltreatment died from intentional maltreatment-specific injuries at a rate 5.9 times that of unreported children, and they died from accidental injuries at twice the rate.

Extending this research, we examined whether rates of injury death varied by type of prior child maltreatment allegation. Results showed that children with prior allegations of physical abuse sustained fatal injuries at a rate 1.7 times that of children with prior allegations of neglect, and they died from inflicted (or intentional) injuries at a rate 5 times that of those with neglect allegations (Putnam-Hornstein et al., 2013b).

We also used these same data to analyze sleep-related deaths. Results showed that for both sudden unexpected infant death and sudden infant death, a prior referral to CPS was an incredibly strong risk factor relative to other demographic and health fields examined (Putnam-Hornstein et al., 2014; Shapiro-Mendoza et al., 2014). After adjusting for baseline risk factors, the rate of sudden infant death was more than 3 times as great among infants reported for possible maltreatment than for those with no CPS history. As these studies illustrate, birth cohort data provide a unique opportunity to examine the important antecedents of rare events and provide policy makers and practitioners with actionable information for developing risk assessments and targeting services.

3.5.3 Looking at "Old" Questions Through the Lens of "New" Fields

The rich data fields available in birth records offer ways to explore existing child welfare questions through new lenses. Studying racial and ethnic disparities in child welfare involvement is one such example. In California, data consistently show a pronounced overrepresentation of Black children in the CPS system relative to their White counterparts. One prominent hypothesis has always been that these disparities are largely a function of poverty. Hispanic children, however, who have similar rates of poverty as Black children, have rates of CPS involvement closer to those of White children. We used birth cohort data to better understand these contradictions.

We explored population-level racial disparities often observed at the different points of child welfare involvement (referral, substantiation, and entry to foster care) for all children born in 2002 by race and ethnicity using enhanced data from birth records (Putnam-Hornstein et al., 2013c). Specifically, using the maternal birthplace field from the birth record, we subdivided Hispanic children into those born to native-born parents (born in the United States) and those born to foreign-born parents (children of immigrants). Results showed that Hispanic children who had native-born parents had rates of child welfare involvement comparable to Black children, whereas children of immigrant parents had very low rates of child welfare contact. After adjusting for socioeconomic and health indicators, the relative risk of referral, substantiation, and foster care entry was significantly lower for both Black and Hispanic children (regardless of maternal nativity) compared to socioeconomically similar White children.

Not only did the birth cohort approach allow for examination of population-level rates of child welfare contact, but the additional covariates available in the birth data provided explanatory power regarding an issue that had long stumped researchers.

3.5.4 Connecting to Other Health Surveillance Programs and Systems

Although administrative data from multiple sources have become increasingly accessible to researchers, at times the data required to answer specific questions remain difficult to access or the fields required for linkage do not exist. In particular, data from health surveillance systems and programs, which are often governed by more stringent privacy rules, can be difficult to obtain for linkage. In these cases, birth records, which are more universally linked to other data sources, have proved a useful bridge between otherwise unlinked data systems.

For example, in studying child maltreatment outcomes in California, we have been working to link child welfare event data to emergency department and hospitalization records maintained by the state. Although these data were not available for direct linkage at the time of this publication, we connected the programs through their common linkage to birth records. Specifically, the state's Maternal Health Surveillance Program links birth records to hospitalization data to examine perinatal conditions of women who are discharged from hospitals. Using existing record linkages between child welfare and birth data, we linked child welfare and hospitalization data through birth records.

By exploiting a common point of connection between two datasets, we developed studies that examined child maltreatment risk associated with various health conditions of both the mother and child. For example, using diagnostic codes from hospital discharge records, we followed all children born in 2006 to examine levels of child welfare system contact associated with different forms of substance exposure at birth (Prindle et al., 2018). Prenatal substance exposure was strongly associated with an infant's likelihood of being reported to CPS by age 1. Significant variation in the likelihood and level of CPS involvement was observed by substance exposure type.

We also used these same 2006 birth cohort data to examine the presence of mental health disorders among mothers and the timing and relationship of child welfare involvement for their children (Hammond et al., 2017). Results showed that 34.6% of infants born to mothers with a diagnosed psychiatric disorder were reported to CPS within 1 year of birth, and many of those reports were made within the first month of life (77.2%). In contrast, among children born to mothers without a mental health disorder, only 4.4% were reported to CPS during this same period.

Using birth records as a common linkage key between child welfare and health data allowed us to develop important new research regarding the child maltreatment risk associated with substance exposure and mental health disorders.

3.5.5 Documenting Intergenerational Maltreatment and CPS Involvement

Linked birth cohort data have been instrumental in documenting intergenerational maltreatment dynamics. Recently, a team of CDN researchers used multiple years of vital birth records to publish studies examining child welfare risk among children born to teens with histories of substantiated child maltreatment (Cederbaum et al., 2013; Putnam-Hornstein et al., 2013a, 2015). Although the period differed, in each study, birth records for infants born to teen mothers during a specific period were extracted from California vital birth records. Maternal information from the birth record was then linked to historical CPS data to identify young mothers with histories of substantiated maltreatment prior to giving birth. Linkages to CPS data were also completed for their infants to prospectively identify child victims.

First, the methodology was used to establish rates of abuse and neglect history among adolescent mothers between the ages of 12 and 19 in 2009. We found that before pregnancy, 44.9% of mothers had been reported as alleged victims of maltreatment, 20.8% had been substantiated as victims, and 9.7% had spent time in foster care (Putnam-Hornstein et al., 2013a).

We then examined child maltreatment risk for infants born to a similar population of teen mothers (aged 15 to 19) in 2006 and 2007 (Putnam-Hornstein et al., 2015). Nearly one-quarter (23.6%) of infants were reported to CPS for possible abuse or neglect and 7.8% were substantiated as victims before age 5. The level of maternal maltreatment history was significantly related to children's maltreatment risk. Among teen mothers with no maltreatment, 17.4% of their children had reports of maltreatment made to CPS before age 5, compared to 35.9% for children of mothers with histories of unsubstantiated reports of maltreatment and 44.1% for children of mothers who had been substantiated as victims of abuse or neglect.

Using birth cohort data, the research showed that children born to adolescent mothers without any history of child welfare involvement had rates of child welfare involvement similar to children of non-teen mothers. Much of the enhanced risk for teen mothers, therefore, relates to child welfare involvement—the trauma inflicted by the abuse and neglect or instability and family disruption that may have occurred.

We utilized a similar methodology to examine health disparities among children born to adolescent mothers between 2007 and 2009 and found a small increased risk of low birth weight among infants whose mothers had histories of child maltreatment (Cederbaum et al., 2013).

The results highlight the richness of the birth data. With information on as many as three individuals, the method can be used to conduct family-level research on child maltreatment risk.

3.5.6 Using "Big" Data to Look at Small, Understudied Groups

Last, administrative data is a powerful tool to study hard-to-reach or underrepresented populations and sensitive or stigmatized topics that are often difficult to examine using surveys (Brownell & Jutte, 2013; Christensen, 1958; Harron et al., 2017). When birth data are included in linkages, this power increases.

Using birth data, we expanded our work regarding differences in child maltreatment rates among children of foreign-born versus US-born mothers to look at this phenomenon among smaller racial and ethnic groups and by country of maternal origin (Finno-Velasquez et al., 2017; Johnson-Motoyama et al., 2015). For example, one group often understudied by child welfare researchers is children of Asian or Pacific Islander descent. This group of children has lower overall rates of CPS involvement compared to other races and ethnicities. However, when we disaggregated the group by maternal birth location, consistent with our earlier study of Hispanic children, we found very similar rates between children with foreign-born mothers versus US-born mothers. Although Asian and Pacific Islander populations are often combined due to their smaller size, our results revealed substantial heterogeneity in child maltreatment risk among these subgroups.

When we examined child maltreatment reports by age 5 for Asians and Pacific Islanders by country of maternal origin, we found additional important differences. For example, approximately 2% to 4% of children born to Chinese or Indian mothers had a CPS report. In contrast, roughly 20% of children born to Pacific Islander, Hawaiian, or Samoan mothers had CPS contact. Without the substantial power that the large birth cohort provides, establishing such estimates for these smaller population groups would not be possible.

3.5.7 Community Monitoring of Risks and Assets at Birth

Birth data can also be used for community monitoring of risks and assets at birth. Currently, we are working with various counties in California to consider how we can begin using birth cohort data to standardize and monitor both factors. Using this information, we can adjust the number of service delivery slots required in our communities and neighborhoods to address the needs of children with a greater concentration of risks. For example, using terrain maps, we have explored areas with the greatest density of births and compared them to areas with the greatest density of children who have the highest risk of maltreatment referrals, to identify areas in need of prevention resources. Finally, we have used linked birth data to support more informed program delivery and better targeting of resources to at-risk communities.

3.6 Conclusion

Vital birth records are one of the most underutilized sources of administrative data. Their oversight is unfortunate given the tremendous amount of information that is collected at birth. Birth records are universally collected and nationally standardized; contain rich and detailed health, demographic, financial, and service information on multiple individuals; and can serve as a population base (spine) for developing prospective studies regarding children.

This chapter has reviewed select examples of innovative child maltreatment research made possible using birth records as the anchor for linking administrative records. These advances include the ability to calculate cumulative rates of child welfare involvement; develop population-based prospective studies; look at "old" questions through the lens of "new" fields; connect to other health surveillance programs and systems; document intergenerational maltreatment and child welfare involvement; use "big" data to look at small, understudied groups; monitor risks and assets at birth for communities; and apply risk stratification for service delivery and maltreatment prevention.

Admittedly, researchers can access birth data more readily in some states than in others. This inconsistency should be addressed. Researchers have an opportunity to come together to advocate for federal clarification on how birth records can be used. Greater access to this data resource across the country would open opportunities to develop cross-state projects and replicate some of the innovative work that has been done so far. Ultimately, greater utilization of birth record data can help us ask and answer important questions about child maltreatment and develop policies and programs to better serve children and families at risk.

References

Abraham, K. G., Haskins, R., Glied, S., Groves, R. M., Hahn, R., Hoynes, H., Liebman, J. B., Meyer, B. D., Ohm, P., Potok, N., Rice Mosier, K., Shea, R. J., Sweeney, L., Troske, K. R., & Wallin, K. R. (2017). The promise of evidence-based policymaking: Report of the Commission on Evidence-Based Policymaking. https://www2.census.gov/adrm/fesac/2017-12-15/Abraham-CEP-final-report.pdf

Brownell, M. D., & Jutte, D. P. (2013). Administrative data linkage as a tool for child maltreatment research. *Child Abuse Neglect, 37*(2–3), 120. https://doi.org/10.1016/j.chiabu.2012.09.013

Card, D. E., Chetty, R., Feldstein, M. S., & Saez, E. (2010). Expanding access to administrative data for research in the United States. *SSRN Electronic Journal*. https://doi.org/10.2139/ssrn.1888586

Cederbaum, J. A., Putnam-Hornstein, E., King, B., Gilbert, K., & Needell, B. (2013). Infant birth weight and maltreatment of adolescent mothers. *American Journal of Preventive Medicine, 45*(2), 197–201. https://doi.org/10.1016/j.amepre.2013.03.016

Christensen, H. T. (1958). The method of record linkage applied to family data. *Marriage and Family Living, 20*(1), 38–43. https://doi.org/10.2307/347362

Connelly, R., Playford, C. J., Gayle, V., & Dibben, C. (2016). The role of administrative data in the big data revolution in social science research. *Social Science Research, 59*, 1–12. https://doi.org/10.1016/j.ssresearch.2016.04.015

Culhane, D. P., & Metraux, S. (1997). *Where to from here? A policy research agenda based on the analysis of administrative data*. https://repository.upenn.edu/cgi/viewcontent.cgi?article=1086&context=spp_papers.

Dunn, H. L. (1946). Record linkage. *American Journal of Public Health, 36*(12), 1412–1416. https://doi.org/10.2105/AJPH.36.12.1412

Enamorado, T., Fifield, B., & Imai, K. (2019). Using a probabilistic model to assist merging of large-scale administrative records. *American Political Science Review, 113*(2), 353–371. https://doi.org/10.1017/S0003055418000783

Finno-Velasquez, M., Palmer, L., Prindle, J., Tam, C. C., & Putnam-Hornstein, E. (2017). A birth cohort study of Asian and Pacific Islander children reported for abuse or neglect by maternal nativity and ethnic origin. *Child Abuse & Neglect, 72*, 54–65. https://doi.org/10.1016/j.chiabu.2017.07.009

Hammond, I., Eastman, A. L., Leventhal, J. M., & Putnam-Hornstein, E. (2017). Maternal mental health disorders and reports to child protective services: A birth cohort study. *International Journal of Environmental Research and Public Health, 14*(11), 1320. https://doi.org/10.3390/ijerph14111320

Harron, K. (2016). *Introduction to data linkage*. Administration Data Research Network https://adrn.ac.uk/media/174200/data_linkage_katieharron_2016.pdf

Harron, K., Dibben, C., Boyd, J., Hjern, A., Azimaee, M., Barreto, M. L., & Goldstein, H. (2017). Challenges in administrative data linkage for research. *Big Data & Society, 4*(2), 205395171774567. https://doi.org/10.1177/2053951717745678

Hotz, V. J., Goerge, R., Balzekas, J., & Margolin, F. (Eds.). (1998). *Administrative data for policy-relevant research: Assessment of current utility and recommendations for development*. Joint Center for Poverty Research.

Johnson-Motoyama, M., Putnam-Hornstein, E., Dettlaff, A. J., Zhao, K., Finno-Velasquez, M., & Needell, B. (2015). Disparities in reported and substantiated infant maltreatment by maternal Hispanic origin and nativity: A birth cohort study. *Maternal and Child Health Journal, 19*(5), 958–968. https://doi.org/10.1007/s10995-014-1594-9

Kim, H., Wildeman, C., Jonson-Reid, M., & Drake, B. (2017). Lifetime prevalence of investigating child maltreatment among US children. *American Journal of Public Health, 107*(2), 274–280. https://doi.org/10.2105/AJPH.2016.303545

Meyer, B. D., Mok, W. K. C., & Sullivan, J. X. (2015). Household surveys in crisis. *Journal of Economic Perspectives, 29*(4), 199–226. https://doi.org/10.1257/jep.29.4.199

National Center for Health Statistics. (2019). *National Vital Statistics System: Revisions of the U.S. standard certificates and reports*. https://www.cdc.gov/nchs/nvss/revisions-of-the-us-standard-certificates-and-reports.htm

National Research Council. (2009). Uses of vital statistics data. In M. J. Siri & D. L. Cork (Eds.), *Vital statistics: Summary of a workshop* (pp. 9–34). National Academies Press.

O'Hara, A., Shattuck, R. M., & Goerge, R. M. (2017). Linking federal surveys with administrative data to improve research on families. *The Annals of the American Academy of Political and Social Science, 669*(1), 63–74. https://doi.org/10.1177/0002716216678391

Prindle, J. J., Hammond, I., & Putnam-Hornstein, E. (2018). Prenatal substance exposure diagnosed at birth and infant involvement with child protective services. *Child Abuse & Neglect, 76*, 75–83. https://doi.org/10.1016/j.chiabu.2017.10.002

Putnam-Hornstein, E. (2011). Report of maltreatment as a risk factor for injury death: A prospective birth cohort study. *Child Maltreatment, 16*(3), 163–174. https://doi.org/10.1177/1077559511411179

Putnam-Hornstein, E., Cederbaum, J. A., King, B., Cleveland, J., & Needell, B. (2013a). A population-based examination of maltreatment history among adolescent mothers in California. *Journal of Adolescent Health, 53*(6), 794–797. https://doi.org/10.1016/j.jadohealth.2013.08.004

Putnam-Hornstein, E., Cleves, M. A., Licht, R., & Needell, B. (2013b). Risk of fatal injury in young children following abuse allegations: Evidence from a prospective, population-based study. *American Journal of Public Health, 103*(10), e39–e44. https://doi.org/10.2105/AJPH.2013.301516

Putnam-Hornstein, E., Needell, B., King, B., & Johnson-Motoyama, M. (2013c). Racial and ethnic disparities: A population-based examination of risk factors for involvement with child protective services. *Child Abuse & Neglect, 37*(1), 33–46. https://doi.org/10.1016/j.chiabu.2012.08.005

Putnam-Hornstein, E., Schneiderman, J. U., Cleves, M. A., Magruder, J., & Krous, H. F. (2014). A prospective study of sudden unexpected infant death after reported maltreatment. *The Journal of Pediatrics, 164*(1), 142–148. https://doi.org/10.1016/j.jpeds.2013.08.073

Putnam-Hornstein, E., Cederbaum, J. A., King, B., Eastman, A. L., & Trickett, P. K. (2015). A population-level and longitudinal study of adolescent mothers and intergenerational maltreatment. *American Journal of Epidemiology, 181*(7), 496–503. https://doi.org/10.1093/aje/kwu321

Putnam-Hornstein, E., Ahn, E., Prindle, J. J., Magruder, J., Webster, D., & Wildeman, C. (2021). Cumulative rates of child protection involvement and terminations of parental rights in a California birth cohort. *American Journal of Public Health, 111*(6), 1157–1163. https://doi.org/10.2105/AJPH.2021.306214

Roos, L. L., Brownell, M., Lix, L., Roos, N. P., Walld, R., & MacWilliam, L. (2008). From health research to social research: Privacy, methods, approaches. *Social Science & Medicine, 66*(1), 117–129. https://doi.org/10.1016/j.socscimed.2007.08.017

Schwartz, S. (2009). The U.S. Vital Statistics System: The role of state and local health departments. In M. J. Siri & D. L. Cork (Eds.), *Vital statistics: Summary of a workshop* (pp. 77–86). National Academies Press.

Shapiro-Mendoza, C. K., Camperlengo, L., Ludvigsen, R., Cottengim, C., Anderson, R. N., Andrew, T., Covington, T., Hauck, F. R., Kemp, J., & MacDorman, M. (2014). Classification system for the Sudden Unexpected Infant Death Case Registry and its application. *Pediatrics, 134*(1), e210–e219. https://doi.org/10.1542/peds.2014-0180

Simon, A. (2014). Using administrative data for constructing sampling frames and replacing data collected through surveys. *International Journal of Social Research Methodology, 17*(3), 185–196. https://doi.org/10.1080/13645579.2012.733176

Victor, T. W., & Mera, R. M. (2001). Record linkage of health care insurance claims. *Journal of the American Medical Informatics Association, 8*(3), 281–288. https://doi.org/10.1136/jamia.2001.0080281

Chapter 4
Going Beyond Where You Live: Innovative Uses for Spatial Data Using Linked Child Welfare Datasets

Bridget Freisthler, Nancy Jo Kepple, and Jennifer Price Wolf

In this chapter, we examine possible ways to use linked data that go beyond the resources available and risks experienced in those neighborhoods where people live. Specifically, we focus on whether those resources and risks differ if you consider where a person goes or spends time through their daily living activities. To address these issues, we need to understand how neighborhood effects have traditionally been measured, the new approaches to understanding contextual effects, and how linked data may include vital address information, which details important information about where a person spends time.

4.1 Traditional Approach to Understanding Neighborhood Effects

Traditional approaches to understanding risks for families at the ecological level are place-based and focus on residential neighborhoods when identifying and addressing problems like child maltreatment. In administrative databases, we may have access to the address where a person lives as it provides the information needed to conduct the investigation of child abuse and neglect. The home addresses in these databases can be geocoded (i.e., create x, y coordinates to place on a map) and aggregated to

B. Freisthler (✉)
College of Social Work, Ohio State University, Columbus, OH, USA
e-mail: freisthler.19@osu.edu

N. J. Kepple
School of Social Welfare, University of Kansas, Lawrence, KS, USA

J. Price Wolf
School of Social Work, San Jose State University, San Jose, CA, USA
e-mail: Jennifer.wolf@sjsu.edu

C. M. Connell, D. M. Crowley (eds.), *Strengthening Child Safety and Well-Being Through Integrated Data Solutions*, Child Maltreatment Solutions Network, https://doi.org/10.1007/978-3-031-36608-6_4

provide rates of maltreatment across geographic areas, such as zip codes or Census tracts.

This approach can be useful because it identifies specific geographic areas that are extremely of high risk, such as those that are and have been entrenched in problematic parenting practices over time. However, it may not adequately address overall rates of maltreatment if the geographic areas identified are too small, or it may dilute our ability to determine high-risk neighborhoods if the geographic areas identified are too big.

In Fig. 4.1a, we present a map of rates of referrals, using administrative units at the Census-tract level. Rates of referrals are based on the underlying population of children within each Census tract in 2001. Figure 4.1b shows the same areas for 2008. Over time, overall rates of referrals have decreased, except that those areas with high rates of referrals continue to have higher rates of referrals relative to other areas. This pattern indicates that these areas may benefit from a place-based community intervention. Implementing interventions using a place-based strategy limits intervention activity to specific geographic areas, making it more feasible to deliver or implement. On the other hand, using Census tracts to depict a "neighborhood" applies arbitrary administrative boundaries that may or may not reflect what a person views as their "neighborhood" or accurately represents those contexts where a person spends their time.

If you consider a typical day, parents are likely to spend time in a range of places, such as the grocery store, their place of employment, their child(ren)'s school, or other such places (e.g., a coffee shop or gym). Similarly, if you consider the range of daily activities in which an individual participates, some of those activities occur outside of the specific location where they live (i.e., their "residential neighborhood"). This has implications for how we study the effects of "neighborhoods" on health and social problems, such as child abuse and neglect. To better understand how a range of environments might affect parenting norms or behaviors, we can study the differential effect of residential neighborhood vs. activity spaces on behaviors.

4.2 Importance of Understanding Environmental Exposure

Before describing additional ways to measure contexts, we discuss how exposure to various environments may negatively or positively affect behaviors and social problems. To begin, environmental exposures may influence behaviors through both direct and indirect mechanisms. The literature on contextual cueing provides some insight about the influence of regular, passive exposure. Because of mechanisms that help to focus attention, individuals are only consciously aware of a small amount of the information they visually process. That being said, attention can be attuned to specific stimuli based on regular exposure and implicit learning. Contextual cueing allows the brain to focus attention on aspects of the environment in a non-random way by prioritizing cues with significant behavioral relevance (Brockmole et al., 2006; Chun & Jiang, 1998). In fact, we can shape the larger

Fig. 4.1 Rate of referrals for child welfare investigations per 1000 children by Block Group in Sacramento, California. (**a**) Rate of referrals by Block Group in 2001. (**b**) Rate of referrals by Block Group in 2008

environmental context to guide attention to specific targets within a complex array of information (Chun & Jiang, 1998). Thus, what an individual sees on a regular basis may affect their behaviors, particularly, the ways in which the environment is shaped to highlight specific elements of the place where they are. For example, an individual may see alcohol advertisements on the way home that cue them to purchase alcohol or even seek out a specific bar after a stressful day. Marketing researchers know this: in order to increase the likelihood of responding to an advertisement, individuals often need to be exposed to a visual, tangible cue multiple times (Lane & Fastoso, 2016; McCoy et al., 2017).

Cueing that occurs while a person goes about their daily activities and routines may be a mechanism through which alcohol advertising or seeking alcohol outlets affect drinking behaviors. After a stressful day at work, a parent may pass the store where they regularly purchase alcohol and make an impromptu stop to make a purchase that increases the likelihood of more frequent and/or heavy drinking behaviors. In turn, drinking alcohol is related to increased use of punitive parenting practices, placing a child at risk of physical abuse (Berger, 2005; Freisthler & Gruenewald, 2013). Alternatively, a parent may consciously travel to specific contexts for socializing and drinking that increase drinking expectancies and are not within their home and/or neighborhood context (Monk & Heim, 2013). Excluding contexts that explicitly cue drinking expectancies could limit our understanding of parental exposure to drinking opportunities. If contextual cueing is a factor that influences drinking, understanding the spaces where parents interact with these cueing messages could help provide information about environmental risk.

4.3 Looking Beyond Neighborhood Context

Activity spaces, which measure mobility across environments, may be one such tool. Activity spaces are environments where parents spend time, such as at work, at their child's school, or with friends and family. In addition, activity spaces may also provide information about social interactions that arise within these spaces. This type of information can expand our understanding of how connection to social supports occurs (or does not occur) across physical and social spaces and how these shape parenting behaviors.

Having small activity spaces may indicate a person has limited mobility for any number of reasons. For example, it may indicate that they are not able to drive in a city with little public transportation or they require the use of a wheelchair, walker, or baby stroller that makes it difficult to move around one's environment when sidewalks are nonexistent. Limited English skills could also make interaction in the larger community difficult. As a result, some individuals may have very limited mobility, which could conscribe their ability to seek out diverse resources or facilitate social connections across multiple networks. This, in turn, could limit the number of "weak ties" in their social network, which provide access to several different forms of human and social capital (Granovetter, 1977). In addition to weak ties, limited mobility could impact connection to supports that provide tangible and emotional resources, if these supports are geographically distant. However, connectivity and access to informal supports may be facilitated by proximity if social supports exist within one's activity spaces. Thus, activity spaces that measure those locations a parent is likely to go on a daily or regular basis may provide a better measure of opportunities to access resources or to be affected by risky environments than just looking at where they live.

In light of these ways that environmental context may shape behaviors, both activity patterns (defined by the paths people travel and operationalized by networks/roadways/linear distance) and activity spaces (defined by the surrounding environment to which they are exposed and operationalized by polygons/areas) may better identify true exposures to resources and risks found in the areas in which people actually spend their time compared to the perspective obtained when only considering a person's residential address. As people move throughout their day, they do so in various contexts that occur in and out of their residential neighborhood. Each location an individual visits has the potential to expose them to helpful resources and/or to risks for participating in harmful behaviors. For example, a grocery store may have access to healthy and nutritious foods, but it likely also stocks unhealthy foods, cigarettes, and alcohol—all of which can pose a variety of risks for individuals. Prior work has demonstrated that we cannot assume individuals will always patronize stores that are closest to their home; rather they may stop by places that are more convenient along their activity patterns or within their activity spaces, such as the store proximal to their place of employment (Sastry et al., 2002). We need to think more comprehensively about the places people go to more accurately model exposure to resources and risk.

Despite the fact that these activities do not always occur within the neighborhood area where a person lives, when examining the person-in-environment, we have historically assessed the resources and risks available in residential neighborhoods. As one example, public health practitioners regularly show maps that depict mortality rates across a city and describe how the zip code where one lives may also affect how long one lives. While this map may be used to exemplify social determinants of health (e.g., how poverty, lack of access to health care, and racial inequality may affect lifespan), it is assumed that the effects of environmental exposures are uniform across neighborhoods. In fact, the strength of these environmental exposures could vary based on factors such as the level of residential instability and the extent to which residents remain primarily within their own neighborhood. Residential neighborhoods may consequently be an accurate measurement if one examines risks and resources unique to where someone lives, but could miss other potential exposures that could raise or lower risks of problematic behaviors.

In Fig. 4.2, we present the daily activity path of a person that depicts where she spent time during one 24-hour period. This figure shows that parents are likely to visit places that have a range of child abuse and neglect rates while conducting their daily life activities. In this particular example, the parent lives in a place where the rate of substantiations is a little higher than the rate for the county. However, where

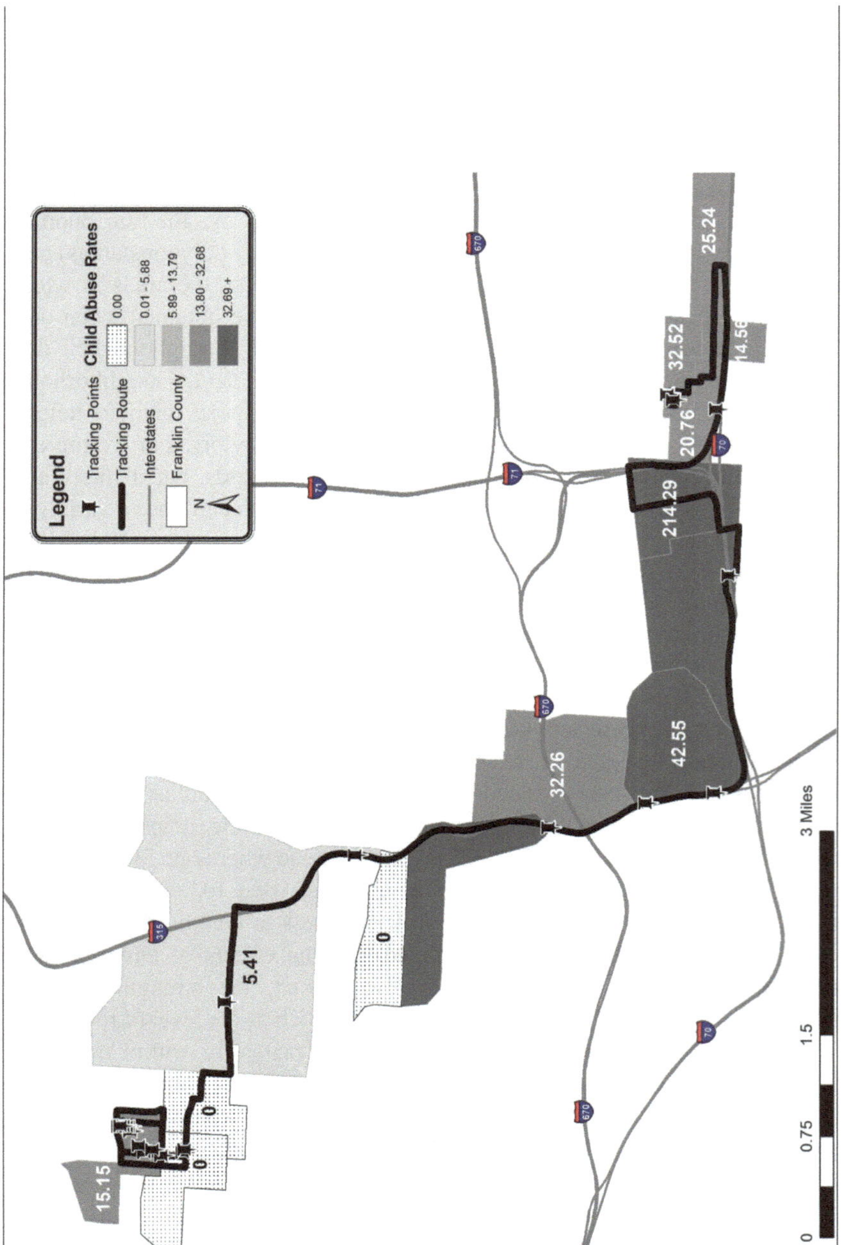

Fig. 4.2 Daily travel pattern for a person overlaid onto rates of substantiated child abuse (per 1000 children) by census tract

she works, the rate is 2.6 times the rate of maltreatment for the county and more than double that of where she lives. *Do we really know which environment is most likely to influence parenting practices?*

What if a residential neighborhood is relatively devoid of risks or of resources? Or what if those risk or resources look different from what we typically measure? Much of the research on neighborhood context as it relates to child abuse and neglect has been conducted in primarily urban areas (cf. Freisthler et al., 2006; Coulton et al., 2007). Do these same risks and resources translate to the same social problems when we move outside of cities? As one example, rural areas may be high in poverty (a risk factor) but have close social connections where everyone provides support when needed (a resource). Do these social connections buffer the effects of poverty in these rural areas when material resources are low? Alternatively, does social isolation exacerbate risk in high-need families within rural contexts where social connections may be a predominant resource?

Examining residential neighborhoods has provided us with important information on how place may affect child maltreatment rates. As this field of research advances, focusing on ways that better capture where parents and children spend time may provide a more nuanced understanding of the effects of place on abusive and neglectful parenting.

4.4 New Approaches to Understanding the Role of Place on Child Maltreatment

Other scholarly disciplines (e.g., urban planning, geography) are moving away from studying residential neighborhoods to focus on activity spaces instead. Prior work in these fields has highlighted how activity spaces are unique to each individual and reflect those places where people spend time. Examining activity spaces, as opposed to residential neighborhoods, could vastly change how we characterize the risks a person is exposed to or the resources available to them.

Using exposure to alcohol as an example, residential neighborhoods assume that alcohol outlets located in the area we live are likely to indicate the exposure to opportunities to drink (Freisthler et al., 2014). The underlying theory argues that more alcohol outlets provide more opportunities to drink and subsequently more alcohol-related problems. This argument has also been extended to activity spaces. Freisthler et al. (2014) present the activity space (measured using a convex hull polygon, the smallest convex polygon able to contain all of the coordinates of a person's activity space) and residential neighborhood for two individuals living in the same Census tract. They show that comparing densities of alcohol outlets in activity spaces meant that one parent could be exposed to over 100 alcohol outlets during their daily activities, while the other parent was only exposed to five. However, only looking at residential neighborhoods means both people would have an exposure of five outlets. This likely underestimates the true exposure of outlets for the first parent. Importantly, the same activity space map could represent resources, such as child-friendly venues like parks or social service agencies, as well.

In Fig. 4.3, we show the home addresses of two respondents who participated in a survey about family life. Their residential neighborhoods, as measured by Census tract (see Fig. 4.3a, b), only represent a very small portion of where they spend time. The stars represent the regular/routine places these parents stated they went during the past week. A visual examination of these locations makes it apparent that the activity spaces of each person's daily life looks very different. One individual has very clustered activity locations, while the other's activity locations are much more diffused.

We can measure activity spaces in a variety of ways. Survey-based activity spaces often distill an activity space to some of the most common places people are likely to go. In the case of parents, this may include home, work, children's schools, day cares, grocery stores, other retail stores, pharmacies, parks or playgrounds, or where a child plays sports. To obtain this information, we might ask for an exact street address, which includes a house number and street name, or we might ask for the street the place is located and the nearest cross street. These addresses are then geocoded so that we can create activity spaces. Below, we discuss the various ways to depict activity spaces using survey data.

In Fig. 4.3c, d, we show destination nodes for our two respondents. Essentially, these are the Census tracts where all of those activities occur. Nodes are used under the assumption that people tend to "bundle" activities. For example, individuals are likely to go to the gas station near where they go shopping or near their work location. As shown, in some cases the Census tracts are fairly small and likely a good representation of where the parent is spending time. In others, the Census tract is actually much larger than where the parent spends time. This depiction of activity space allows one to continue to use many administrative data sets that are available to examine the environmental context of where a parent spends time.

Activity spaces identified through shortest network distance can be found in Fig. 4.3e, f. Shortest network distance is the shortest path between all of the points using the local roadway networks. We limit the network by having it begin at the home but let the end points vary based on the shortest path between all the points. As with destination nodes, people are likely to bundle activities and errands along their regular travel patterns. Generally, a buffer (e.g., 500 meters) is drawn around the network in order to create density measures. We should note that when we use this algorithm, we do not know whether this is actually the route the person takes to go to these places. The algorithm used can be changed to account for public transportation routes and could be ordered based on the sequence of locations a person visits. One limitation for this measure is that, for the person in Fig. 4.3f who regularly attends football and baseball games (as noted by the star in the furthest location), it is difficult to fully estimate the effect of the travel patterns on possible risks and resources. In other words, without using GPS monitoring, we do not know how he or she gets to the location of the stadium—using regularly traveled routes, public transportation, or a less commonly traveled route.

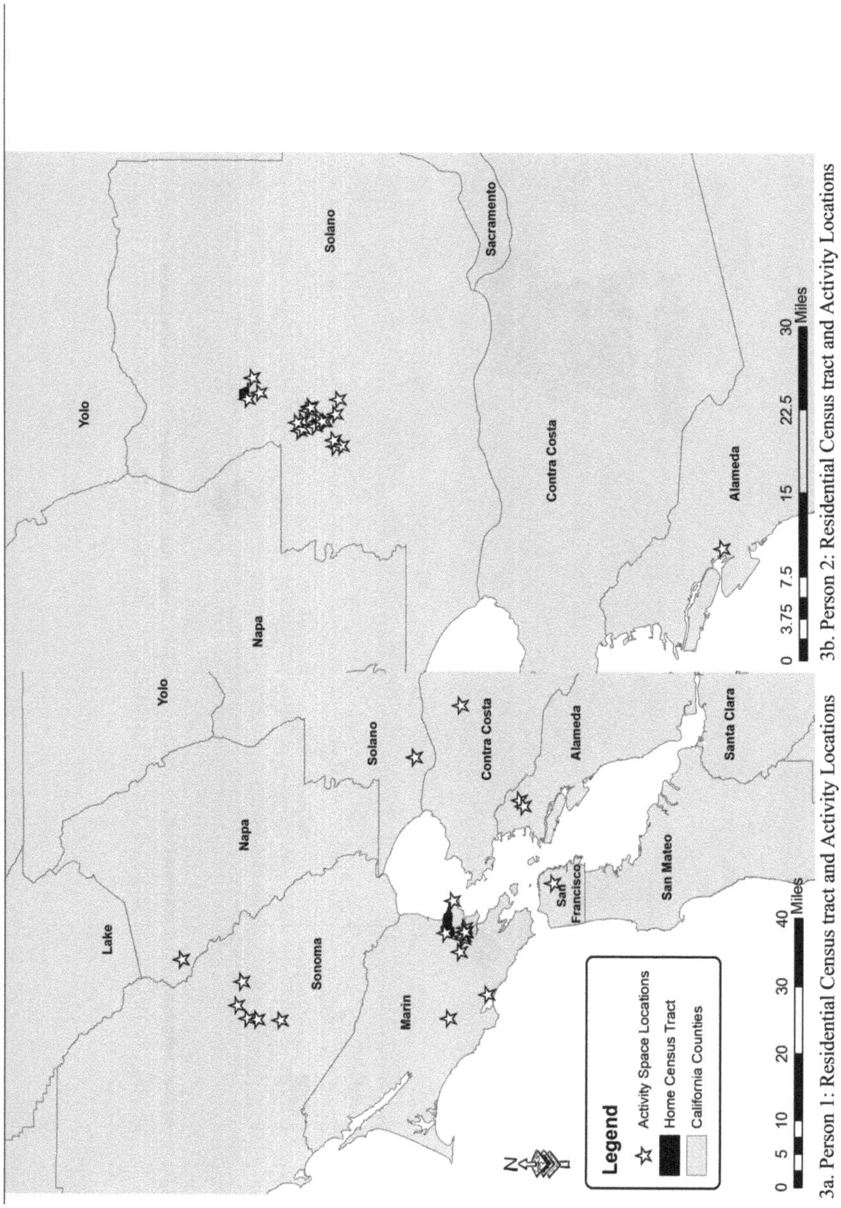

Fig. 4.3 Depictions of residential neighborhood, activity locations, activity patterns, and activity spaces for two people. (**a**) Person 1: residential census tract and activity locations. (**b**) Person 2: residential census tract and activity locations. (**c**) Person 1: destination nodes for activity locations. (**d**) Person 2: destination nodes for activity locations. (**e**). Person 1: shortest network distance activity pattern. (**f**) Person 2: shortest network distance activity pattern. (**g**) Person 1: activity spaces using 1 and 2 standard deviational ellipses and convex hull polygon. (**h**) Person 2: activity spaces using 1 and 2 standard deviational ellipses and convex hull polygon

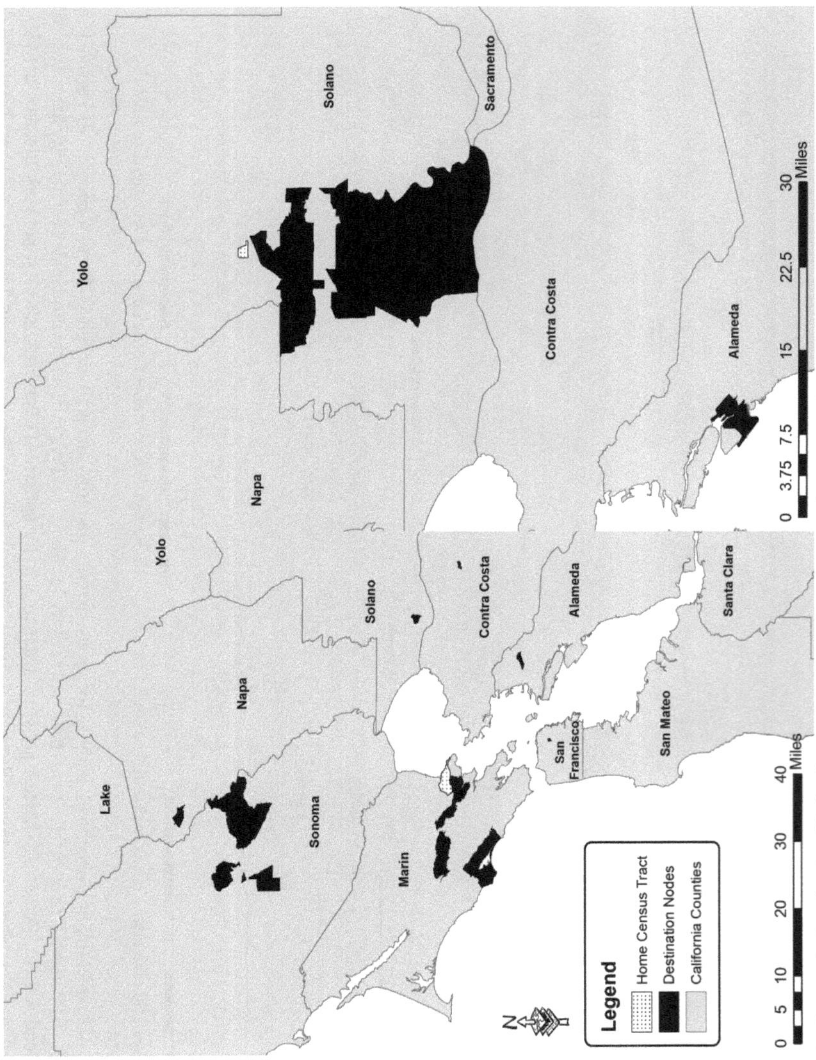

3c. Person 1: Destination Nodes for Activity Locations

3d. Person 2: Destination Nodes for Activity Locations

Fig. 4.3 (continued)

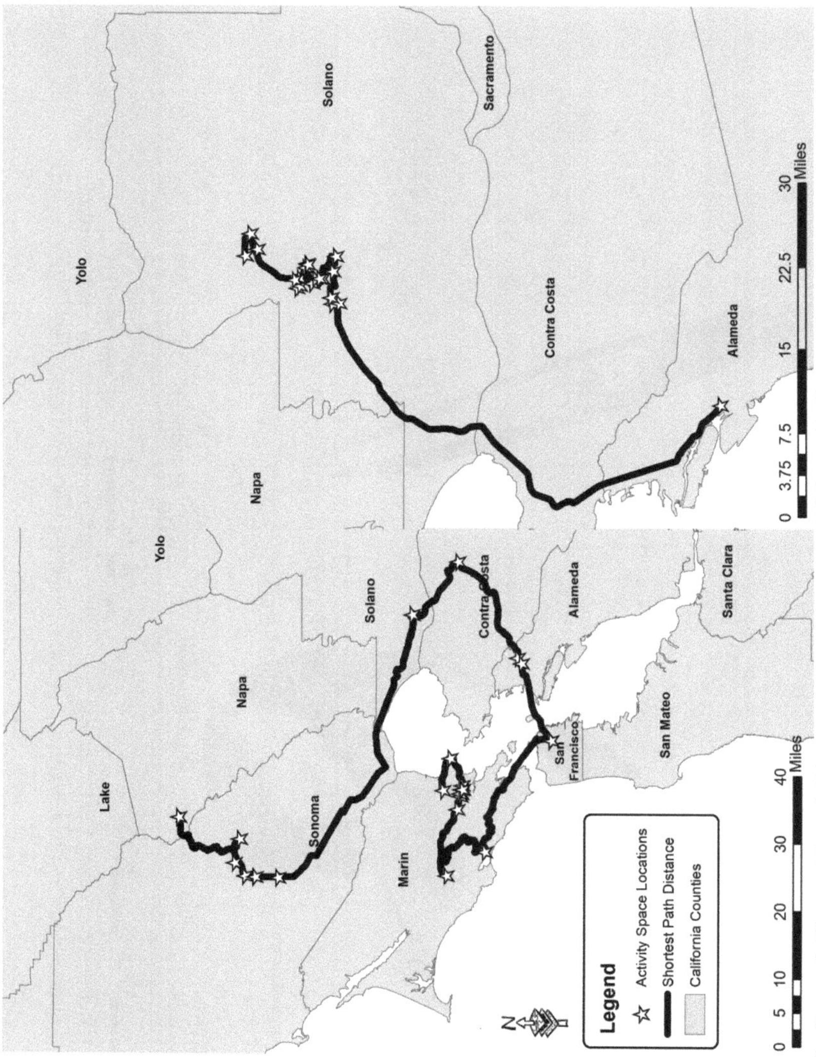

3e. Person 1: Shortest Network Distance Activity Pattern

3f. Person 2: Shortest Network Distance Activity Pattern

Fig. 4.3 (continued)

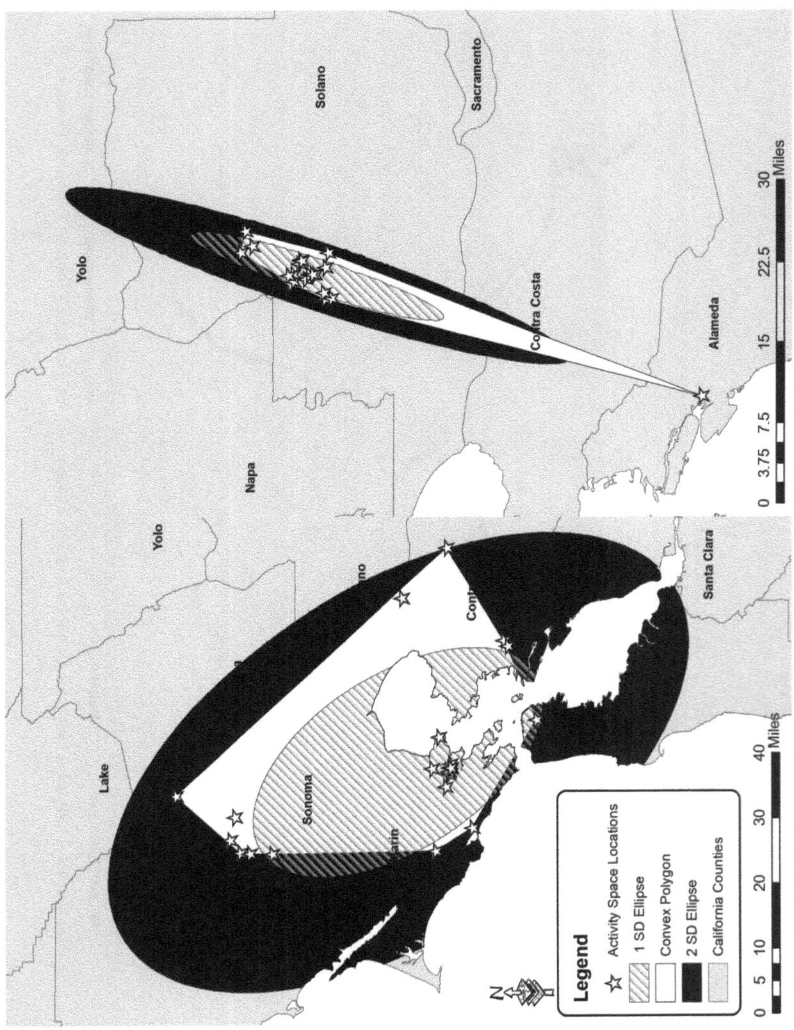

3g. Person 1: Activity Spaces using 1 and 2 Standard Deviational Ellipses and Convex Hull Polygon

3h. Person 2: Activity Spaces using 1 and 2 Standard Deviational Ellipses and Convex Hull Polygon

Fig. 4.3 (continued)

We will now move from point-based approaches to polygons that seek to encompass an area defined by the locations of the points. In one approach, activity spaces can be represented by standard deviational ellipses. These ellipses cover a percentage of the points based on whether they are one or two standard deviations away in the areal field. Figure 4.3g, h show the one standard deviation ellipse in gray stripes, which encapsulates about 67% of the points for the average activity space. For the more diffuse activity points in Fig. 4.3g, this means that several points fall outside of the polygon, while in Fig. 4.3h, the stadium "point" is not included. The ellipse in black represents a two standard deviation ellipse; an ellipse that covers, on average, 95% of activity points. Again, for the more diffuse activity points in Fig. 4.3g, this means that several points continue to fall outside of the polygon. On Fig. 4.3h, the same number of points is included; it is just the size of the ellipse that differs. Finally, we show the convex hull polygon activity spaces using white color in Fig. 4.3g, h. This polygon uses the shortest distance to encompass all of the points included as part of the activity space. This is one of the most common depictions of activity spaces in social science (non-geography) literature. These types of polygons are useful for showing the range of areas a person may visit through their daily activities; however, it can be difficult to use these spaces with many of the administrative data sets we use to describe environmental context.

Although surveys are useful when obtaining information from a large number of people, this approach limits the number of locations one can identify. Another limitation is that not all locations can be geocoded, as respondents are not always able to provide us addresses or cross streets reliably. This can affect the size and shape of the activity space polygon. To create a convex hull polygon, we need to have at least three useable locations, so missing or incorrect address information could result in a lot of missing data. An alternative to a survey-based approach is to use GPS monitoring to identify a person's activity spaces.

Using GPS monitoring, a device (e.g., GPS-enabled cell phone, Fitbit™) can be used to ping a person's location at specific intervals and provide x, y coordinates for mapping. These GPS devices also provide a much better indication of where a person spends time and for how long. To illustrate what we might learn using GPS-based activity spaces, Fig. 4.4 presents a map of Franklin County, Ohio, and the substantiated rates of child abuse and neglect by Census block group. Overlaid onto this map are the 24-hour GPS activity patterns for two individuals. The figure illustrates that throughout the day, they are spending time in "neighborhoods" with a wide range of child abuse and neglect rates. What might this mean for parenting? If we think parenting is transmitted through social norms and networks (see Freisthler et al., 2016), is the place that a parent lives the most relevant location for understanding where parenting behaviors have been learned or honed?

Fig. 4.4 Rates of substantiated child abuse and neglect per 1000 children by census tract and two individuals tracking points and routes

4.5 Using Linked Data to Create Activity Spaces

Linked data may be one way to create activity spaces without the expense of conducting population-based surveys or downloading enormous amounts of geographic data from GPS-enabled devices. Activity spaces may be useful for identifying which service providers fall within areas where a person regularly spends time, helping to ensure that a parent can more easily make appointments. If we know the activity spaces of siblings not placed together, we may also be able to identify better locations for joint visits to maintain those connections.

Administrative data is a veritable treasure trove when it comes to address information. Child welfare records can tell us the location of where a child and parent lived at the time of the abuse or neglect and the foster care, group home, or residential placement addresses if the child is out-of-home. For children, additional information can be found with linked education data. School records allow us to track where children are going to school. Early education or day care locations may be available for younger children participating in federally funded programs like Head Start. Medicaid records may tell us about where a child or parent received mental health or substance misuse services (if Medicaid-billable) and locations of doctors' offices that they visit. Electronic health records may provide better information about locations of doctors' offices when parents and children do not have Medicaid.

These records could be used to develop activity spaces across the life span. Electronic health records will have dates of visits and where the child was living at the time. School records will also show what schools a child attended each year. Thus, yearly activity spaces for children and parents could be created to assess long-term exposures to risk and resources.

4.6 Activity Spaces and Parenting: What We Know

To date, very little research has examined how activity spaces are related to parenting behaviors, including child abuse and neglect. We conducted a small study in Los Angeles, where we looked at parenting behaviors in relation to the size of activity space. What we found was the larger the activity spaces, the less the parent used punitive discipline practices (Freisthler et al., 2016). We concluded that smaller activity spaces may be indicative of parents who are more isolated, a known risk factor for child maltreatment.

However, as we argued in our 2016 paper, what such findings may reveal is how social norms of parenting are transmitted across spaces. If we view parenting behaviors as "contagious," like the measles or the whooping cough, a larger activity space may lead to exposure to more individuals who use the current "normative" parenting behaviors. In the USA, these normative behaviors tend to be non-punitive types of parenting. Thus, having a larger activity space leads to exposure to more types of parenting behaviors and people with different parenting norms. For parents who engage in punitive parenting practices, they may face having their actions corrected in public when using those practices.

Herd immunity might be another public health analogy that could describe how activity spaces may affect parenting. Does going to places where non-punitive parenting is practiced by more people result in "herd immunity" that inoculates families who may be struggling in other ways? If so, the benefits would actually be the spread of positive parenting norms through more exposure to more non-punitive types of parenting behaviors or guardians who will censure punitive parenting practices.

Before we go too much into detail about activity spaces, do we even know whether risks (or resources) differ when we look at residential neighborhoods vs. activity spaces? In Freisthler et al. (2021), we show that the density of alcohol outlets in residential neighborhoods is not correlated with any of the measures of activity spaces or patterns described here. Thus, relying solely on residential neighborhoods may incorrectly assess resources available for parents or the risks experienced. However, we did find that parents who used corporal punishment or child physical abuse clustered at the same stores, not in the same residential neighborhoods (Freisthler et al., 2020). As we move forward, assessing how activity spaces and different measurements of activity spaces are related to parenting behaviors might enable us to better identify how parents use space or how we can modify public places to improve parenting behaviors.

4.7 Challenges and Opportunities in Creating Activity Spaces with Administrative Data

One of the biggest challenges we face in using linked data is understanding how address data are stored. For example, when individuals attend their yearly doctor's visit, the receptionist will often confirm that person's address. If that individual has recently moved, they will be asked to provide updated information. Noting a change, the receptionist may delete the old address from the records and enter the updated address. In some systems, such updates may be maintained as historical information, but in others the updated information overwrites the previous record. If administrative records write over address fields and do not save the historical records, it makes it much harder to recreate historical or lifetime activity spaces.

Among the first questions to ask when looking into using addresses in linked data are as follows: When were those data collected (e.g., intake, last known address)? How often they are updated? Are historical addresses saved or deleted? If you receive data downloads going back in time, can you retrieve that historical data?

Another challenge is the lack of geographic information system (GIS) ability among practitioners, social service agencies, and social work educators. For example, some child welfare agencies automatically add x, y coordinates when addresses are entered. However, these data are rarely used, nor are there plans to use them in a systematic way. Part of the reason we do not mine geographic data fully may be that we do not provide training in their use. Very few social work programs provide GIS as a regular part of their curriculum. Nor do we teach our doctoral students how to conduct analyses correctly using spatial data. Thus, GIS and spatial data analysis are seen as a black box—too difficult to master or not relevant to the day-to-day practice of a social worker.

Despite these challenges, activity spaces hold promise for identifying how the environment affects parenting and risk for child maltreatment and are an area rich for study. These activity spaces may be developmental, such that parents of young children may have smaller activity spaces than those with older children. As younger children are more likely to experience abuse or neglect, further exploration of this topic could lead to interventions that seek to expand activity spaces for new parents. The activity spaces here are relatively crude in the sense that we seek to identify the most common locations where parents spend time. A more sophisticated activity space measure would weigh the risks and resources of the locations where parents go based on the amount of time spent at those locations. Finally, given the issues related to disproportionality in the child welfare system, how might segregation affect activity spaces and availability of resources? Given the resources involved in survey and GIS data-collection methods, linked spatial data could advance this area of inquiry more quickly and lead to novel place-based solutions that could reduce abuse and neglect.

Acknowledgments This project was supported by grant number P60-AA-006282 from the National Institute on Alcohol Abuse and Alcoholism. The content is solely the responsibility of the authors and does not necessarily represent the National Institute on Alcohol Abuse and Alcoholism or the National Institutes of Health.

A version of this chapter was presented by the first author at the Strengthening Child Safety and Wellbeing through Integrated Data Solutions Conference in Pennsylvania State University on September 27, 2018.

References

Berger, L. M. (2005). Income, family characteristics, & physical violence toward children. *Child Abuse & Neglect, 29*, 107–133.

Brockmole, J. R., Castelhano, M. S., & Henderson, J. M. (2006). Contextual cueing in naturalistic scenes: Global and local contexts. *Journal of Experimental Psychology: Learning, Memory, and Cognition, 32*(4), 699–706.

Chun, M. M., & Jiang, Y. (1998). Contextual cueing: Implicit learning and memory of visual context guides spatial attention. *Cognitive Psychology, 36*, 28–71.

Coulton, C., Crampton, D., Irwin, M., Spilbury, J., & Korbin, J. (2007). How neighborhoods influence child maltreatment: A review of the literature and alternative pathways. *Child Abuse & Neglect, 31*, 1117–1142.

Freisthler, B., & Gruenewald, P. J. (2013). Where the individual meets the ecological: A study of parent drinking patterns, alcohol outlets and child physical abuse. *Alcoholism: Clinical and Experimental Research, 27*(6), 993–1000.

Freisthler, B., Merritt, D. H., & LaScala, E. A. (2006). Understanding the ecology of child maltreatment: A review of the literature and directions for future research. *Child Maltreatment, 11*(3), 263–280.

Freisthler, B., Lipperman-Kreda, S., Bersamin, M., & Gruenewald, P. J. (2014). Tracking the when, where, and with whom of alcohol use: Integrating ecological momentary assessment and geospatial data to examine risk for alcohol-related problems. *Alcohol Research: Current Reviews, 36*, 29–38.

Freisthler, B., Thomas, C. A., Curry, S. R., & Price Wolf, J. (2016). An alternative to residential neighborhoods: An exploratory study of how activity spaces and perception of neighborhood social processes relate to maladaptive parenting. *Child and Youth Care Forum, 45*(2), 259–277. https://doi.org/10.1007/s10566-015-9329-7

Freisthler, B., Thurston, H., & Price Wolf, J. (2020). An exploratory study of parenting in public places: What can we learn from parents' activity locations and physical punishment? *International Journal on Child Maltreatment: Research, Policy and Practice, 3*(2), 249–269.

Freisthler, B., Kepple, N. J., Wolf, J. P., & Carson, L. (2021). Activity spaces: Assessing differences in alcohol exposures and alcohol use for parents. *GeoJournal, 86*(1), 69–79.

Granovetter, M. S. (1977). The strength of weak ties. In *Social networks* (pp. 347–367). Academic Press.

Lane, V. R., & Fastoso, F. (2016). The impact of repeated ad exposure on spillover from low fit extensions to a global brand. *International Marketing Review, 33*(2), 298–318.

McCoy, S., Everard, A., Galletta, D. F., & Moody, G. D. (2017). Here we go again! The impact of website ad repetition on recall, intrusiveness, attitudes, and site revisit intentions. *Information & Management, 54*(1), 14–24.

Monk, R. L., & Heim, D. (2013). Panoramic projection: Affording a wider view on contextual influences on alcohol-related cognitions. *Experimental and Clinical Psychopharmacology, 21*(1), 1.

Sastry, N., Pebley, A. R., & Zonta, M. (2002). *Neighborhood definitions and the spatial dimension of daily life in Los Angeles* (CCPR Working Paper 033–04). California Center for Population Research, University of California–Los Angeles.

Chapter 5
Leveraging Harmonized Multi-System Administrative Data to Examine Experiences and Outcomes for Child Protective Services-Involved Children, Youth, and Families

Lawrence M. Berger

The United States spends billions of dollars each year on a wide range of social welfare programs funded, administered, and delivered at the federal, state, and local levels by a multitude of agencies. Enormous amounts of data are collected in the process of administering and delivering these programs. These data have immense

The Wisconsin Administrative Data Core was developed by the Institute for Research on Poverty (IRP) at the University of Wisconsin-Madison, in collaboration with a range of State of Wisconsin agencies over multiple decades and benefitted greatly from the leadership of Maria Cancian and Jennifer Noyes. A large number of IRP and State agency staff were and are involved in this ongoing process. Unfortunately, I am unable to acknowledge them each individually. However, I and the IRP community are grateful to staff and leadership at the Wisconsin Departments of Children and Families, Health Services, Workforce Development, Public Instruction, and Corrections for their ongoing partnerships, facilitation of data access, and consultation and collaboration on research questions and priorities, as well as policy and data issues. These partnerships have been crucial to the development, sustainability, and usefulness of the Data Core. The following IRP staff have played key roles in designing, constructing, and managing the Data Core: Pat Brown, Steve Cook, Dan Ross, Hilary Shager, Jane Smith, Katie Thornton, Maggie Darby Townsend, and Lynn Wimer. I am also grateful to my collaborators in the studies summarized in this chapter: Maria Cancian, Michael Collins, Laura Cuesta, Sarah Font, Leah Gjertson, Eunhee Han, Jennifer Noyes, Vanessa Rios-Salas, Kristi Slack, and Tim Smeeding. The research presented here was supported by funding from the Eunice Kennedy Shriver National Institute of Child Health and Human Development (R21 HD091459) and the Wisconsin Department of Children and Families through research agreements with IRP including the "Child Support Policy Research Agreement" and the "Wisconsin Educational Collaboration for Youth in Foster Care" project, funded under the US Department of Health and Human Services Child Welfare–Education System Collaborations to Increase Educational Stability grant program. Additional support was provided by IRP. The agencies involved do not certify the accuracy of the analyses presented.

L. M. Berger (✉)
University of Wisconsin-Madison, Madison, WI, USA
e-mail: lmberger@wisc.edu

C. M. Connell, D. M. Crowley (eds.), *Strengthening Child Safety and Well-Being Through Integrated Data Solutions*, Child Maltreatment Solutions Network, https://doi.org/10.1007/978-3-031-36608-6_5

potential to inform evidence-based decision-making and to enable better policy and program design, monitoring, and evaluation. Yet, they are vastly underutilized and their accessibility to researchers, policymakers, and practitioners is quite limited. This is, in part, due to the complexities involved in harmonizing and linking data from multiple agencies (potentially at multiple levels of government) with varying mandates, using different administrative data systems, collecting data for different purposes, and subject to different rules governing privacy, data sharing, and data utilization. At the same time, we are in the midst of an exciting and dynamic era for considering innovative administrative data harmonization, access, and utilization initiatives, and related researcher-practitioner partnerships, as exemplified by the final report of the bipartisan Commission on Evidence-Based Policymaking (2017) and the Foundations for Evidence-Based Policymaking Act of 2018, as well as government, academic, and philanthropic leaders' calls for major new initiatives using administrative data for policy research and effective governance.

Child protective services (CPS) research, policy, and practice stand to benefit substantially from such initiatives. CPS represents an expansive and expensive array of policies, programs, and services intended to respond to (risk of) child abuse and neglect. It is a far-reaching system that disproportionately affects socially and economically disadvantaged children and families—particularly families of color—the majority of which are involved in other social welfare programs typically outside the purview of CPS. In 2017, for example, approximately 5.6% of all US children (7.5 million) were investigated by CPS and nearly 1% (674,000) were determined to be victims of child maltreatment (U.S. Department of Health and Human Services, 2019). Approximately 685,000 US children spent some time in foster care (U.S. Department of Health and Human Services, 2018). Yet, these annual prevalence rates substantially underestimate CPS involvement over the course of childhood. Between birth and age 18, 37% of all children and 53% of black children are investigated by CPS (Kim et al., 2017), 13% of all children and 21% of black children are determined by CPS to be victims of abuse or neglect (Wildeman et al., 2014), and 6% of all children and 12% of black children spend some time in foster care (Wildeman & Emanuel, 2014). Moreover, estimates from Wisconsin indicate that more than 80% of CPS-involved families are participating in one or more other social welfare programs, such as the Supplemental Nutrition Assistance Program (SNAP; formerly Food Stamps) or Medicaid (Berger, 2017; Cancian et al., 2017a). CPS is also expensive, with public spending, including that on foster care, reaching nearly $30 billion per year (Rosinsky & Williams, 2018).

Long-standing collaborations between researchers at the University of Wisconsin-Madison Institute for Research on Poverty (IRP) and leadership in multiple Wisconsin state agencies have resulted in the development of an extensive longitudinal linked administrative data system, known as the Wisconsin Administrative Data Core, that supports integrated analyses of earnings, income, multiple program participation trajectories, including CPS involvement, and the well-being of individuals and families. These collaborations and the resulting Data Core stand to provide a national model for state-level data integration and utilization efforts in the social welfare arena. In this chapter, I first describe the Wisconsin

Administrative Data Core and its development, highlighting key implications for similar efforts. I then summarize several recent projects in the child welfare arena that utilized these integrated data.

5.1 The Wisconsin Administrative Data Core: History and Development

The Wisconsin Administrative Data Core consists of cleaned, harmonized, and linked data from multiple administrative data systems across government agencies. It has been developed and maintained by the IRP in collaboration and through data sharing agreements with its Wisconsin state agency partners. The Data Core is designed to support integrated analyses of earnings, income, and multiple program participation trajectories of families in Wisconsin, as well as health, social, educational, and economic well-being for these families. The data allow for longitudinal, cross-program analyses that would not be possible using a single agency's data. It was developed over multiple decades, initially on a project-based, ad-hoc basis. For each project, data linkage and harmonization began with a sample drawn from a single data system. Sample members were then linked to their records in other administrative data systems to create a specialized multi-system extract. For a detailed history of the Data Core, its construction, and its evolution, see Brown et al. (2020). A description of the Data Core is available at https://www.irp.wisc.edu/wadc/. This section draws extensively from both.

Since 2008, the Data Core has been created on an annual basis in comprehensive fashion, so as to include the entire population of individuals in each available data system. Its initial construction was enabled by funding from the Office of Planning, Research, and Evaluation (OPRE), Administration for Children and Families (ACF), and US Department of Health and Human Services under the *Federal-State Partnerships to Build Capacity in the Use of TANF and Related Administrative Data* for the project "Building an Integrated Data System to Support the Management and Evaluation of Integrated Services for TANF-Eligible Families" (Principal Investigator: Maria Cancian). It is currently fully funded by research grants and contracts from state and federal agencies and private foundations to IRP investigators.

The Data Core is created each year by linking data from the full universe of participants in any of the State of Wisconsin's electronically available social welfare administrative data systems to create a single file of unique individuals (including adults and children) participating in one or more of the included programs, achieved through a combined process of deterministic, probabilistic, and "chaining" matching methods. Matching identifiers include name, sex, Social Security number (SSN) and SSN verification code, cross-system personal identification numbers (program-specific IDs), dates of birth and death, places of birth, parent identifiers (names, dates of birth, SSNs of mother and father), race/ethnicity, indicators for members of multiple births (twin, etc.), and indicators for adoption. However, the identifiers differ in availability across data systems. Because not all cases are deterministically matched across all of the systems, it is possible that some individuals are incorrectly matched or have duplicate records. As such, the data are not appropriate for case management, monitoring, or enforcement, and the IRP does not provide matched individual or case-level data back to state agencies.

The core data file, known as the Multi-Sample Person File (MSPF), includes a single record for each individual that appears in the administrative data of any of the contributing state programs. To construct the MSPF, IRP programming staff access fully identifiable data from each of the contributing programs under data-sharing agreements with the relevant agencies. The MSPF is rebuilt annually using the most recent data from each system. Annual reconstruction allows for ongoing improvements in matching and harmonization, as well as inclusion of new records. Data from all programs/systems in which the individual appears—across the entire period in which data are available—are incorporated in the individual's record. Demographic and residential location information for each individual from the date of first observation through most recent observation is also included in the MSPF. The current version of the MSPF includes more than 8.2 million individuals.

Beyond providing a single multi-system record for each individual, the MSPF allows individuals to be observed across systems over time. It also enables linkages among parents and children, members of the same family or household (to the extent these relationships can be determined from the available data), and individuals comprising a benefit case, resulting in data files that include multi-system administrative data on multiple individuals with familial or residential ties followed across time. In addition to the MSPF, the Data Core includes group aggregation files that identify individuals in families (using various definitions) and members of the same benefit cases. It also includes program participation files with information about eligibility, participation, service spells, and benefit receipt and amount by specific time period, as well as cross-walk files that can be used (through masked identifiers that match identifiers in other samples) for linkages to other data systems not included in the MSPF. The data can also be geocoded and thereby linked to contextual data. All files are linkable by unique, IRP-constructed IDs.

The fully linked data systems used to build the MSPF are provided by the Wisconsin Departments of Children and Families, Health Services, Public Instruction, Workforce Development, and Corrections, as well as the Milwaukee County Sheriff's Office. The data include:

- SNAP and its predecessor, Food Stamps, participation and benefits, for which consistent data are available from 1989 to present
- BadgerCare (Wisconsin's combined Medicaid and State Child Health Insurance program) participation from 1989 to present and claims and encounters from 2009 to present
- Temporary Assistance for Needy Families and its predecessor, Aid to Families with Dependent Children, participation and benefits from 1989 to present
- Child Support Enforcement participation (as payee or payer), orders, payments (paid or received), and arrears from 1996 to present

- Child Care Subsidy participation and benefits from 1997 to present
- Unemployment Insurance (UI) wages from 1988 to present and benefits from 2007 to present
- State prison incarceration from 1990 to present
- Milwaukee County Jail incarceration from 1994 to present (more limited data are available in some earlier years)
- CPS involvement from 2004 to present (with more limited data available beginning in 1990)
- Public school records from the 2005–2006 school year to present

Note that wages (earnings) reported to the Unemployment Insurance system, which are available from 1988 to present, are linked to the individuals in the MSPF on an annual basis; however, these records are linked only to individuals who are already included in the MSPF based on participation in another program; the MSPF does not include the universe of individuals with earnings reported to the Unemployment Insurance system. Additional data systems, including the Department of Revenue (tax records), Adult and Juvenile Circuit Court, Family Court, Supplemental Security Income, and Vital Records, have been linked to MSPF data on an ad-hoc basis. The Data Core has also been linked to survey data.

The Data Core is explicitly intended to support research to inform public policy design, administration, and evaluation. That is, it is intended to support actionable, applied work with clear implications for policy and practice, and which addresses the questions that policymakers and practitioners need answered, thereby contributing to evidence-informed policymaking. Its sustainability fully depends on close, collaborative relationships among the IRP and the Wisconsin agencies from which the data are provided. Collaborative relationships are maintained, in part, through regular meetings and briefings, sustained and ongoing research agreements, bidirectional technical assistance provision, project-specific contracts, and quarterly "learning exchanges" bringing together IRP leadership and affiliates with agency leadership and staff. The IRP and its state partners approach the identification of projects such that the agencies propose to IRP projects that are of core policy or practice interest to the agencies (and may or may not be of particular interest to IRP researchers) but require IRP researchers' expertise, IRP researchers propose to the state agencies projects that are of particular interest to the researchers—and have core policy or practice implications—but may not be areas of focus for the agencies, or increasingly, IRP and agency partners collaborate to develop questions and projects that are of high interest to both.

It is crucial to recognize that the IRP does not own the data included in the Data Core. Rather, the IRP houses these data on behalf of the state. As such, all projects must be approved by each state agency from which any data will be used. Furthermore, the investigators must provide evidence that the study has potential to inform state policy and/or programs, as state agencies are prohibited from providing data for research that is not relevant to their mission. Projects must also be approved by the University of Wisconsin-Madison Institutional Review Board. Currently, data use is also contingent on an on-campus IRP affiliate being a principal or co-investigator on the project. Because construction and administration of the Data Core require significant resources that are solely provided by users' projects, without any source of independent infrastructure support, projects are charged a data access and use fee based on the breadth and scope of data to be used. Researchers are given access only to data that have identifiers removed and have been cleaned and harmonized for research purposes. Data access is subject to strict confidentiality and security rules, and the data are stored on a HIPAA-compliant server. Researchers can access the data only on this server and are unable to remove any individual-level data. State agencies from which data are used are guaranteed a 30-day review period before any results are presented or published. However, researchers have full academic freedom for presentation or publication after this review.

The Wisconsin experience developing the Data Core offers several insights for future efforts. First, a trusting, collaborative, and mutually beneficial partnership between state agencies and the research institution is crucial. It is also critical to clearly delineate where each of the partners' objectives align and diverge, as well as what each partner will contribute to the common goals. For example, in creating and using the Wisconsin Administrative Data Core, the state agencies provide data and substantive expertise on both the programs and data systems, including how individual fields are used in practice, whereas the IRP provides specialized programming staff, technical support staff, hardware and software, and methodological expertise for analyses. All of the organizations provide one another with ongoing technical assistance and are committed to open and frequent communication.

There are other mutual benefits of the collaboration. For example, the state agencies benefit from a fuller understanding of the range of programs with which their clients are engaged, but have neither the mandate nor the capacity to take on large, cross-system research endeavors. IRP researchers benefit from access to unprecedented population data, which enable them to pursue core policy-relevant research into the causes and consequences of poverty and inequality to inform programs to reduce them, in some cases through experimental or quasi-/natural-experimental methods. In addition, because the IRP does not own the data, nor use them for program or case monitoring, it has the flexibility to engage in probabilistic matching, which is not afforded to state agencies. This is crucial given that most administrative systems are not generally designed with a strict requirement to maintain a single record per individual and, across administrative systems, there is no commonly mandated set of high-quality identifiers. Also, agencies and systems vary (for legitimate reasons tied to their specific mandates) in the range and quality of demographic variables collected, and individual agencies with distinct missions face challenges in determining which data source takes precedence and how inconsistencies in the data should be resolved. Agencies' priorities are to privilege those data that are most critical to their mission, rather than cross-system consistency. The IRP is able to resolve many such difficulties—for research purposes—because it faces less pressure to ensure legally accurate data for a specific individual at a given point in time and instead can prioritize efforts on arriving at scientifically accurate aggregate estimates.

The Wisconsin experience also highlights several considerations for constructing and maintaining an integrated data system, which should be addressed from the onset. Key decisions include whether to use samples or the full universe of individuals in specific systems; whether data should be structured by individual, case, family unit, or child/adult status; what common identifiers are essential; how matching will be conducted (deterministically, probabilistically, or through a combined process), how uncertainty in matching will be addressed, and whether matches must be legally accurate; how administrative data will be pre-processed and cleaned for research purposes; how and when the integrated system will be updated and/or expanded; and how changes in the administrative data system (or any of the systems therein) will be documented, addressed, and reconciled with historical data.

It is also crucial to develop an explicit approval and governance structure to ensure compliance with established statutes and regulations; provide for data security and privacy protection; and guide who has access to the data, under what circumstances, and how, including explicit requirements of the types of questions and projects that are acceptable for data access (e.g., whether studies must explicitly inform policy and programs or may focus exclusively on basic research). Finally, an explicit plan for addressing ongoing infrastructure and resource needs to ensure long-term sustainability should be developed to include partners' roles and responsibilities for obtaining or providing funding, as well as management and administration. Sustaining a data system like the Wisconsin Administrative Data Core requires significant resource commitment as the fixed costs of data harmonization, linking, security, and management are large.

5.2 Recent Child Welfare Research Using the Wisconsin Administrative Data Core

The Wisconsin Administrative Data Core has contributed substantially to knowledge about the children and families involved with CPS, the extent to which they are also involved in other social welfare programs, and their ongoing well-being. For example, integrated data analyses indicate high levels of multi-system participation among CPS-involved families (Berger, 2017; Cancian et al., 2017a). In the year prior to a screened-in (investigated) CPS report, for example, approximately 82% of families participated in one or more means-tested social welfare programs, with 16% participating in one such program, 35% in two programs, 24% in three programs, and 7% in four or more programs. Most commonly, families participated in SNAP (74%) and BadgerCare (71%). The data also revealed relatively low levels of employment and earnings: more than 50% of families had at least one-quarter with no reported earnings in the year before the investigation. At the same time, a much smaller, though still substantial, proportion of families received TANF (19%), childcare subsidies (22%), and unemployment insurance benefits (15%).

Approximately 44% of families had a child support order and 33% received some child support in the year before the investigation (Berger, 2017; Cancian et al., 2017a).

These families also experience high levels of inter- and intra-generational involvement in both CPS and the criminal justice system. Berger et al. (2016) report that 28% of CPS-involved children and 34% of children in out-of-home placement in Milwaukee had a parent who was incarcerated in state prison or Milwaukee County Jail in the same year, whereas 18% of adults incarcerated in state prison or Milwaukee County Jail had a CPS-involved child and 6% a child in out-of-home placement in that year. Youth experiencing CPS involvement as adolescents were also highly likely to experience incarceration. Approximately 29% of those who experienced CPS involvement and 34% of those who experienced out-of-home placement at ages 15–16 subsequently had a period of incarceration in state prison or Milwaukee County Jail between ages 18 and 21. Alternatively, of those with a period of incarceration in state prison or Milwaukee County Jail between ages 18 and 21, 19% had been CPS-involved and 7% experienced out-of-home placement at ages 15–16.

This evidence on patterns of social welfare and criminal justice involvement among CPS-involved children and families illustrates the complex array of services and systems with which they are involved. For policy and practice, it points to the need to recognize the various program requirements and mandates that such families face and consider the extent to which they are consistent or conflicting across programs. Below, I briefly describe five studies that were jointly developed by IRP researchers and state agency partners to further inform policy and programs affecting CPS-involved families.

5.2.1 *Can Increasing the TANF Child Support Pass-Through Reduce CPS Involvement?*

Child support paid to TANF recipients is typically subject to state withholding to recover the cost of TANF participation, with a limited pass-through to the custodial parent. As part of Wisconsin's welfare reform, following the Personal Responsibility and Work Opportunity Reconciliation Act of 1996, the state was granted a federal waiver to conduct a randomized experiment in which treatment group families received all of the child support paid on their behalf, without a reduction in their TANF benefits, whereas control group families received only a partial pass-through and benefit disregard, representing the greater of $50 or 41% of child support paid on their behalf in a given month. IRP affiliates Maria Cancian and Daniel Meyer and their colleagues evaluated this experiment, known as the Child Support Demonstration Evaluation (CSDE), using multi-system administrative data linked to participant survey data. CSDE results indicated that treatment group families were more likely to establish paternity, to have child support paid on their behalf, and to receive a larger amount of child support, at no additional cost to the government (Cancian et al., 2008). These findings ultimately influenced child support policy in the United States, Canada, and the United Kingdom.

Although analyzing the influence of child support receipt on child maltreatment was not an aspect of the initial evaluation, a growing body of research linking family income to child maltreatment (Berger & Waldfogel, 2011)—in a context of high levels of CPS involvement at high cost to the public—spurred increased interest among researchers, policymakers, and practitioners in Wisconsin (and nationally) in whether there might be a causal link between income and maltreatment. Given the randomized design of the CSDE and the fact that treatment group assignment resulted in increased income as a result of increased child support, the CSDE data presented a rare opportunity to examine the effect of a random increase in family income on CPS involvement. The resulting study leveraged the CSDE data, which was then linked to administrative CPS records, to examine this effect. Findings indicated that families eligible to receive the full amount of child support paid on their behalf were approximately 10% less likely to experience a CPS investigation during the next 2 years than those receiving a partial child support pass-through (Cancian et al., 2013). These results provided additional evidence of the benefits of a full child support pass-through for TANF-receiving families. Indeed, beyond increasing family income, such a policy has the potential to decrease CPS involvement, an intrusive and expensive intervention. More broadly, the results suggest that income support may be an effective approach for reducing child maltreatment.

5.2.2 *Is Home Foreclosure Associated with CPS Involvement?*

The US housing crisis and subsequent Great Recession of 2008 resulted in an alarming increase in home foreclosures. Foreclosures have the potential to not only affect homeowners but also renters (to the extent that their landlords experience foreclosure). Despite a large literature linking both adverse economic shocks (Berger & Waldfogel, 2011) and housing instability (Cowal et al., 2002; Culhane et al., 2003; Font & Warren, 2013) with increased risk of child maltreatment, as well as evidence that foreclosure is adversely associated with adult and child well-being (Bradbury et al., 2013; Fowler et al., 2015; McLaughlin et al., 2012; Pollack & Lynch, 2009), studies had yet to address the potential impacts of foreclosure on CPS involvement at the micro-level. Two existing macro-level studies, however, suggested a potential link, documenting associations of county-level foreclosure rates with increases in hospital admissions for physical abuse and high-risk traumatic brain injuries (Wood et al., 2012), as well as CPS involvement (Frioux et al., 2014). Yet, it cannot be assumed that such macro-level correlations will hold at the individual or household level.

IRP researchers and their partners at the Wisconsin Department of Children and Families shared a common concern that increases in foreclosures in the state may adversely affect vulnerable children and families. This concern motivated a study of the relation between home foreclosure and CPS involvement (Berger et al., 2015b). To examine this relation, IRP programmers linked Wisconsin Administrative Data Core data to administrative data on foreclosure filings from the Wisconsin Consolidated Court Automation Programs data system, thus allowing for analyses of whether a foreclosure filing on a family's address of record (regardless of whether the family owned or rented the property) was associated with CPS involvement over a 2-year period, accounting for various lag and lead periods. Results indicated that experiencing a foreclosure is associated with substantial increase in the probability of CPS involvement, spanning the year prior to the foreclosure filing and the subsequent year. Importantly, however, such risk was greatest in the period before the filing and shortly thereafter, indicating that increased CPS involvement may not be triggered by the foreclosure filing itself, but by limited economic resources (and, potentially, related stress) during the time leading into and through foreclosure proceedings. These results suggest that, while foreclosure may be a marker for increased CPS involvement, risk of such starts well before a foreclosure filing. As such, prevention efforts should seek to identify families experiencing adverse economic shocks, which may place them at risk of both foreclosure and CPS involvement and target them for intervention well before the actual foreclosure filing itself.

5.2.3 How Does Child Support Enforcement Affect Out-of-Home Placement Trajectories?

As discussed above, the majority of CPS-involved families participate in other social welfare benefit programs, and many are involved with the child support enforcement system. Of particular note, when a child is removed from home by CPS, the state (or county) can reassign an existing child support order to partially recoup out-of-home placement expenditures. Furthermore, the family can be issued a new order to pay child support specifically to reimburse for out-of-home placement costs. While child welfare and child support program administrators and practitioners are aware of these policies, the impacts of such actions on out-of-home placement trajectories had not previously been studied.

In Wisconsin, there is considerable county-level variation in the rate at which existing child support orders are reassigned to recoup out-of-home placement costs, as well as in whether new child support orders for such are successfully enacted. Indeed, use of child support orders to offset out-of-home placement costs varies across counties, from being enacted in less than 10% of out-of-home placement cases to about 80% of such cases (Cancian et al., 2017b). Ongoing discussions among IRP affiliates and Wisconsin Department of Children and Families staff produced a joint interest in understanding the effects of variation of child support enforcement actions of either type on children's out-of-home placement outcomes. To analyze these relations, Maria Cancian and colleagues (2017b) utilized the

Wisconsin Administrative Data Core to leverage a natural experiment harnessing substantial unintentional county variation in child support enforcement actions to examine the effect of out-of-home placement-related child support orders on the length of time children spend in out-of-home placement as a result of CPS removal. They find that imposing child support obligations (of either form) on families with a child in out-of-home placement results in substantially longer out-of-home placement spells. Specifically, a $100 increase in monthly child support (re) assignment leads to a nearly 7-month delay in family reunification. Moreover, they document little association between the likelihood that a family receives an out-of-home placement-related child support order and parental earnings, which stands in stark contrast to typical child support enforcement actions, which generally take parental ability to pay into account. Given the low levels of child support collected through these actions, combined with the high cost of out-of-home placement, these results suggest that imposing child support obligations on families with children in out-of-home placement, particularly those who are likely to eventually reunify, makes little fiscal sense.

5.2.4 Does Out-of-Home Placement Adversely Impact Academic Achievement?

An expansive body of research over many decades has documented that children experiencing out-of-home placement fare poorly on a wide range of outcomes throughout the life course (Blome, 1997; Clausen et al., 1998; Doyle, 2007; Mass & Engler, 1971; Shin, 2004). However, a key concern is whether out-of-home placement causes these poor outcomes or is simply a marker of other risk factors, such that the children experiencing out-of-home placement would have had such outcomes regardless of placement itself (Berger et al., 2009). A number of descriptive reports and studies documenting poor educational outcomes for youth experiencing out-of-home placement (Barrat & Berliner, 2013; Fantuzzo & Perlman, 2007; Scherr, 2007; Smithgall et al., 2004; Stone, 2007) generated considerable concern among Wisconsin Department of Children and Families leadership and staff that out-of-home placement may be adversely affecting the children and youth involved. These concerns resulted in the additional engagement of IRP affiliates and Wisconsin Department of Public Instruction leadership and staff to form the "Wisconsin Educational Collaboration for Youth in Foster Care" collaboration, which was funded under the US Department of Health and Human Services Child Welfare–Education System Collaborations to Increase Educational Stability grant program. The initiative sought to better understand the educational trajectories of youth experiencing out-of-home placement and to design better targeted and administered collaborative interventions between the CPS and educations systems.

One aspect of the collaboration consisted of empirical research to understand whether the association of out-of-home placement with poor educational achievement was likely to be causal in nature. To address this question, IRP affiliates leveraged the Wisconsin Administrative Data Core to examine whether reading and math achievement test scores differed among five groups of children and youth, observed from third to eighth grades: (1) those in an out-of-home placement at the time of the test; (2) those in an out-of-home placement during the year before the test but in-home at the time of the test; (3) those in-home at the time of the test, but subsequently removed from home; (4) those investigated by CPS but not removed from home; and (5) those whose families were economically disadvantaged, as indicated by receipt of SNAP benefits, but who did not experience a CPS investigation or out-of-home placement (Berger et al., 2015a, b). Results from a series of regression models, some of which included child-specific fixed effects, that treated conditions (2) through (5) as counterfactuals for condition (1) revealed that out-of-home placement is unlikely to cause poor academic achievement. Rather, all children involved with CPS exhibit consistently low math and reading achievement scores that are similar in magnitude regardless of whether a child experiences out-of-home placement. Furthermore, the results suggested that children who have not yet been removed from home but will experience out-of-home placement in the future are at highest risk of poor achievement. These findings have led the state to reconsider using out-of-home placement as the primary trigger for education-related intervention and, instead, to consider the larger population of CPS-involved children as at risk for poor educational outcomes.

5.2.5 Does Out-of-Home Placement Lead to Teen Motherhood?

Evidence that out-of-home placement is unlikely to be a cause of poor educational achievement generated interest among both IRP researchers and agency leadership in examining whether out-of-home placement is likely causally linked to additional adverse outcomes that are frequently attributed to experiencing placement. In particular, it is well known that youth experiencing out-of-home placement and, more generally, those who experience CPS involvement, are at increased risk for early motherhood, with estimates suggesting that between 18% and 35% of girls experiencing CPS involvement or out-of-home placement have a teen birth (Doyle, 2007; King, 2017).

To gain insight into whether this association is likely to be causal, Font et al. (2019) estimated a series of survival analyses examining risk of teen motherhood using data on the entire population of female youth in Wisconsin who either (1) experienced out-of-home placement, (2) experienced CPS-involvement but never experienced out-of-home placement, or (3) were in an economically disadvantaged family (defined by the family receiving SNAP benefits) but did not experience CPS involvement or out-of-home placement. The models controlled for a range of sociodemographic characteristics of the youth and their families of origin and accounted for the timing of CPS involvement and out-of-home placement vis-à-vis conception. Results indicated that, while both CPS-involved youth and those experiencing out-of-home placement were at considerably higher risk of teen motherhood than youth from economically disadvantaged families that were not involved with CPS, such risk is unlikely to be caused by CPS involvement or out-of-home placement. Indeed, the risk of teen pregnancy was *lower* during out-of-home placement than before placement. Likewise, among CPS-involved female youth who did not experience placement, risk of teen pregnancy was similar or lower after the CPS investigation than before. In short, CPS-involvement and out-of-home placement do not appear to increase risk of teen motherhood; rather, out-of-home placement may even reduce such risk. Additional findings suggested that the reduced risk associated with out-of-home placement persisted after placement ended for youth who exited care to adoption or guardianship, but not for those who reunified with their families of origin. This suggests that, while out-of-home placement may be beneficial for reducing teen motherhood while youth are in out-of-home care, additional supports and services may be needed once they return home.

5.3 Conclusion

The recent work of the Commission on Evidence-Based Policymaking (2017) and the subsequent Foundations for Evidence-Based Policymaking Act of 2018 suggest a burgeoning federal emphasis on increasing access to and utilization of administrative data for applied social science research. Ongoing collaborations between the IRP and State of Wisconsin agencies, which have resulted in the Wisconsin Administrative Data Core, present a national model for partnering to harness multi-system administrative data to inform policy and program design, monitoring, and evaluation. Multi-system data may be particularly important for child welfare research, given extensive participation of CPS-involved families in an array of social welfare programs. Moreover, such data allow for longitudinal tracking of the health and well-being of these children and families, potentially in the context of randomized or quasi-experimental designs. A range of studies by IRP affiliates have demonstrated the benefits of such data for conducting rigorous research with actionable implications for informing policy and practice, both within CPS and across the many systems with which CPS-involved families interact.

To be sure, there are substantial challenges to building such sustainable collaborations and data systems. These span technical aspects of data harmonization and linking across systems with differing identifiers, creating an efficient and effective governance structure, ensuring data security and privacy, balancing various partners' missions and mandates, and obtaining adequate funding to support the data infrastructure. Nonetheless, the Wisconsin experience provides evidence that these barriers can be overcome and that such collaborations can promote a mutually beneficial culture of evidence-informed policymaking.

References

Barrat, V., & Berliner, B. (2013). *The invisible achievement gap, part 1: Education out-comes of students in foster care in California's public schools*. The Center for the Future of Teaching & Learning at WestEd. Available at: https://www.wested.org/wp-content/files_mf/1400283692Invisible_Achievement_Gap_Full_Report.pdf

Berger, L. M. (2017). *Economic resources and child maltreatment: Implications for the child welfare system*. Annual social policy forum keynote address, Annual meeting of the society for social work and research, New Orleans, LA, January 11-15, 2017.

Berger, L. M., & Waldfogel, J. (2011). Economic determinants and consequences of child maltreatment. *OECD social, employment and migration working paper no. 111*. Organization for Economic Cooperation and Development Publishing.

Berger, L. M., Bruch, S. K., Johnson, E. I., James, S., & Rubin, D. (2009). Estimating the "impact" of out-of-home placement on child well-being: Approaching the problem of selection bias. *Child Development, 80*(6), 1856–1876.

Berger, L. M., Cancian, M., Han, E., Noyes, J., & Rios-Salas, V. (2015a). Children's academic achievement and foster care. *Pediatrics, 135*, e109–e116.

Berger, L. M., Collins, J. M., Font, S. A., Gjertson, L., Slack, K. S., & Smeeding, T. (2015b). Home foreclosure and child protective services involvement. *Pediatrics, 136*(2), 299–307.

Berger, L. M., Cancian, M., Cuesta, L., & Noyes, J. (2016). Families at the intersection of the criminal justice and child protective services systems. *ANNALS of the American Academy of Political and Social Science, 665*, 171–194.

Blome, W. W. (1997). What happens to foster kids: Educational experiences of a random sample of foster care youth and a matched group of non-foster care youth. *Child and Adolescent Social Work Journal, 14*(1), 41–53.

Bradbury, K., Burke, M. A., & Triest, R. K. (2013). *The effect of foreclosure on Boston public school student academic performance*. Working paper no. 13-12. Federal Reserve Bank of Boston.

Brown, P. R., Thornton, K, Ross, D., Smith, J. A., Thornton, K., & Wimer, L. (2020). *Technical report on lessons learned in the development of the institute for research on Poverty's Wisconsin Administrative Data Core*. Available at: https://www.irp.wisc.edu/wp/wp-content/uploads/2020/08/TechnicalReport_DataCoreLessons2020.pdf

Cancian, M., Meyer, D. R., & Caspar, E. (2008). Welfare and child support: Complements, not substitutes. *The Journal of the Association for Public Policy Analysis and Management, 27*(2), 354–375.

Cancian, M., Yang, M., & Slack, K. (2013). The effect of child support income on the risk of child maltreatment. *Social Service Review, 87*, 417–437.

Cancian, M., Noyes, J., & Kim, H. (2017a). *Using linked administrative data to understand the dynamics of multiple program participation*. Report to the Wisconsin Department of Children and Families, December, 2017.

Cancian, M., Cook, S., Seki, M., & Lynn Wimer, L. (2017b). Making parents pay: The unintended consequences of charging parents for foster care. *Children and Youth Services Review, 72*, 100–110.

Clausen, J. M., Landsverk, J., Ganger, W., Chadwick, D., & Litrownik, A. (1998). Mental health problems of children in foster care. *Journal of Child and Family Studies, 7*(3), 283–296.

Commission on Evidence-Based Policymaking. (2017). *The promise of evidence-based policymaking: Report of the commission on evidence-based policymaking*. Commission on Evidence-Based Policymaking.

Cowal, K., Shinn, M., Weitzman, B. C., Stojanovic, D., & Labay, L. (2002). Mother–child separations among homeless and housed families receiving public assistance in New York City. *American Journal of Community Psychology, 30*(5), 711–730.

Culhane, J. F., Webb, D., Grim, S., & Metraux, S. (2003). Prevalence of child welfare services involvement among homeless and low-income mothers: A five-year birth cohort study. *Journal of Sociology and Social Welfare, 30*, 79.

Doyle, J. J., Jr. (2007). Child protection and child outcomes: Measuring the effects of foster care. *American Economic Review, 97*(5), 1583–1610.

Fantuzzo, J., & Perlman, S. (2007). The unique impact of out-of-home placement and the mediating effects of child maltreatment and homelessness on early school success. *Children and Youth Services Review, 29*(7), 941–960.

Font, S. A., & Warren, E. J. (2013). Inadequate housing and the child protection system response. *Children and Youth Services Review, 35*(11), 1809–1815.

Font, S. A., Cancian, M., & Berger, L. M. (2019). Prevalence and risk factors for early motherhood among low-income, maltreated, and foster youth. *Demography, 56*, 261–284.

Fowler, K. A., Gladden, R. M., Vagi, K. J., Barnes, J., & Frazier, L. (2015). Increase in suicides associated with home eviction and foreclosure during the US housing crisis: Findings from 16 national violent death reporting system states, 2005–2010. *American Journal of Public Health, 105*(2), 311–316.

Frioux, S., Wood, J. N., Fakeye, O., Luan, X., Localio, R., & Rubin, D. M. (2014). Longitudinal association of county-level economic indicators and child maltreatment incidents. *Maternal and Child Health Journal, 18*(9), 2202–2208.

Kim, H., Wildeman, C., Jonson-Reid, M., & Drake, B. (2017). Lifetime prevalence of investigating child maltreatment among US children. *American Journal of Public Health, 107*, 274–280.

King, B. (2017). First births to maltreated adolescent girls: Differences associated with spending time in foster care. *Child Maltreatment, 22*(2), 145–157.

Maas, H. S., & Engler, R. E. (1971). *Children in need of parents*. Columbia University Press.

McLaughlin, K. A., Nandi, A., Keyes, K. M., Uddin, M., Aiello, A. E., Galea, S., & Koenen, K. C. (2012). Home foreclosure and risk of psychiatric morbidity during the recent financial crisis. *Psychological Medicine, 42*(7), 1441–1448.

Pollack, C. E., & Lynch, J. (2009). Health status of people undergoing foreclosure in the Philadelphia region. *American Journal of Public Health, 99*(10), 1833–1839.

Rosinsky, K., & Williams, S. C. (2018). Child welfare financing SFY 2016: A survey of federal, state, and local expenditures. *Child Trends*. Available at: https://www.childtrends.org/wp-content/uploads/2018/12/CWFSReportSFY2016_ChildTrends_December2018.pdf

Scherr, T. G. (2007). Educational experiences of children in foster care: Meta-analyses of special education, retention and discipline rates. *School Psychology International, 28*(4), 419–436.

Shin, S. H. (2004). Developmental outcomes of vulnerable youth in the child welfare system. *Journal of Human Behavior in the Social Environment, 9*(1–2), 39–56.

Smithgall, C., Gladden, R. M., Howard, E., Goerge, R., & Courtney, M. (2004). *Educational experiences of children in out-of-home care*. Chapin Hall Center for Children, University of Chicago. Available at: http://citeseerx.ist.psu.edu/viewdoc/download;jsessionid=FCCCA45AEB7F88F15F965F29247D9960?doi=10.1.1.727.5481&rep=rep1&type=pdf

Stone, S. (2007). Child maltreatment, out-of-home placement and academic vulnerability: A fifteen-year review of evidence and future directions. *Children and Youth Services Review, 29*(2), 139–161.

U.S. Department of Health and Human Services, Administration for Children and Families, Administration on Children, Youth and Families, Children's Bureau. (2018). *The AFCARS Report: Preliminary FY 2017 Estimates as of August 10, 2018 – No. 25*. Available from: https://www.acf.hhs.gov/sites/default/files/cb/afcarsreport25.pdf

U.S. Department of Health and Human Services, Administration for Children and Families, Administration on Children, Youth and Families, Children's Bureau. (2019). *Child Maltreatment 2017*. Available from: https://www.acf.hhs.gov/cb/research-data-technology/statistics-research/child-maltreatment

Wildeman, C., & Emanuel, N. (2014). Cumulative risks of foster care placement for American children, 2000-2011. *PLoS One, 9*, e92785.

Wildeman, C., Emanuel, N., Leventhal, J. M., Putnam-Hornstein, E., Waldfogel, J., & Lee, H. (2014). The prevalence of confirmed maltreatment among US children, 2004-2011. *JAMA Pediatrics, 168*, 706–713.

Wood, J. N., Medina, S. P., Feudtner, C., Luan, X., Localio, R., Fieldston, E. S., & Rubin, D. M. (2012). Local macroeconomic trends and hospital admissions for child abuse, 2000–2009. *Pediatrics, 130*(2), e358–e364.

Chapter 6
Too Much, Too Little, or Just Right? How Integrated Data Helps Identify Impact and Opportunity

Melissa Jonson-Reid, Brett Drake, and Maria Gandarilla Ocampo

6.1 Seeking the "Goldilocks Zone" in Child Welfare Policy and Programming

The Goldilocks Zone, in astronomy, is the area around a star where we think that a planet is in the sweet spot to have enough liquid water on it to allow for life, or at least life as we know it (Gary, 2016). We argue there are at least two dimensions of finding the Goldilocks Zone in child protection. The first area of tension might be called "over-intervention vs. under-intervention" and the second could be called "Resource Optimization."

First, the organizations tasked with responding to child maltreatment in this country are constantly trying to understand what is too little, too much, or just right in terms of intervention. In simple terms, we don't want CPS to be too involved and overreact, but we certainly don't want CPS to fail to respond to situations where there is real risk. Society broadly values the autonomy of the family and parental rights and, therefore, asks the child protection system and related family court system to intervene only when necessary to keep a child safe from harm (Zeanah & Humphreys, 2018). While arguments are often made that we undervalue the rights of children to be free of harm and develop (McFall, 2009; Willems, 2019), this continues to be a significant source of tension in our response to abuse and neglect. This tension is further enhanced by additional controversy around perceived differences in surveillance or response according to factors like poverty, racial/ethnic groups, and immigration status. Analyses of integrated data illustrate the substantial promise in determining which families are most likely to benefit from intervention and what level of intervention may be required to produce a positive outcome, while

M. Jonson-Reid (✉) · B. Drake · M. G. Ocampo
Brown School of Social Work, Washington University in St. Louis, St Louis, MO, USA
e-mail: jonsonrd@wustl.edu; brettd@wustl.edu

C. M. Connell, D. M. Crowley (eds.), *Strengthening Child Safety and Well-Being Through Integrated Data Solutions*, Child Maltreatment Solutions Network, https://doi.org/10.1007/978-3-031-36608-6_6

also offering a means of assessing under- or over-intervention of different groups in the population.

The second dimension for child protection has to do with maximizing effectiveness while minimizing cost. This dimension reflects efforts to optimize available resources to best serve children. We do not believe it is farfetched to say that across all forms of social services, there is a constant concern regarding the allocation of scarce resources. Integrated data may help us better target and evaluate services so that we can find our "sweet spot" in child protection. This chapter describes in detail the promise of integrated data in both addressing the "over-intervention vs. under-intervention" problem and the "Resource Optimization" issue.

Integrated data is a commonly used term, but it can look different in different applications. It can include preexisting administrative data from different data sets from different sources that are linked together in an ongoing way. For example, many states are linking child maltreatment, income maintenance, and even data from other sectors (e.g., health, crime) to better evaluate the services that child welfare provides (Chouldechova et al., 2018). Integrated data, however, is not restricted to administrative data alone. Many kinds of data, such as interview data, can be connected to administrative data to create integrated data architecture (Widom, 2017). Integrated data generally refers to data that crosses at least two service systems or organizations, but might refer to large scale multi-system archives like those developed in California (http://cdn.usc.edu/) or Michigan (https://ssw-datalab.org/).

Integrated data, whether purely based on government administrative data or a hybrid mixture of administrative, intervention, and survey data, may help us find the most appropriate approach at the most appropriate intensity with the most appropriate individuals. There are many ways that integrated data can be used to support effective service provision. For instance, these data can be used to:

- Improve our ability to understand client service use
- Identify, engage, and retain clients in treatment
- Characterize treatment as usual or add contextual, explanatory, or control variables
- Understand decision points such as substantiation
- Address issues of effectiveness and cost
- Avoid duplication and verify findings through triangulation

We also argue that by repeating analyses of these types of data over time, we can help identify systemic issues at the policy or community levels that may be hidden in one-off program evaluations.

Of course, linking data across systems can pose challenges. Translation to practice or policy requires considerable effort beyond analyzing data and presenting findings. This chapter considers issues of creating and using integrated data and then effectively translating findings for policy and practice outcomes. We present a case study (drawing on prior work from Stahlschmidt et al., 2018) to illustrate one way these data can be used to identify the population and timing issues and move to translate into services and engagement. We will then consider some existing work to discuss how linked data, in conjunction with outcomes research, can be used to address some of the tensions or fears in regard to equitable targeting and access as well as balance between preserving the family and the child's safety and well-being.

6.2 A Case Example: From Integrated Administrative Data Analysis to Practice Innovation

We present the below case study as an illustrative example of how integrated data can help inform effective practice and policy in Child Protective Services. This work began with a project to link several administrative data sets across agencies to better understand the service paths and outcomes for children reported for abuse or neglect (funded by National Institute of Mental Health: R01 MH061733 & Centers for Disease Control and Prevention: CE001190). In other words, the foundational study's goal was based on following a cohort of children, not created or maintained as an ongoing archive (we will revisit this difference later). The research questions in the study were jointly developed by researchers and state agency leaders who shared a strong history of collaboration. The design of the study involved integrating public child welfare reports and service and placement records with a number of other datasets. These included data from health, income maintenance, law enforcement, special education, juvenile justice, and other systems. The underlying idea was to be able to see how children and families moved through the various systems in order to provide an evidentiary basis for improving policy and service delivery.

Linked data structures like these can be very flexible. For example, such a data set could be used to answer questions about criminal justice outcomes for children contacted by the child welfare system or could be used to determine whether prior parental criminal justice contact might predict child welfare system involvement. The more they grow, the more questions can be answered. One of the key characteristics of the constructed data set was that it included both children contacted by the child welfare system and comparison children of similar socioeconomic status. Accessing income maintenance data allowed us to create separate groups of children who had received Aid to Families with Dependent Children (AFDC) or Temporary Assistance for Needy Families (TANF) but either had or had not had child welfare services system contact. The final result was a data set that included thousands of children that were followed through various administrative datasets for a period of about 20 years (Fig. 6.1)—and that could be used to answer a range of questions.

Building trust between the research team and the involved agencies was of highest importance (Jackson, 2019; Jonson-Reid & Drake, 2008). This involved not only setting up the project, but staying in touch with the data and agency partners, presenting our results, and talking to them about their own questions. For example, one of the early papers developed as a result of this initial trust-building project was

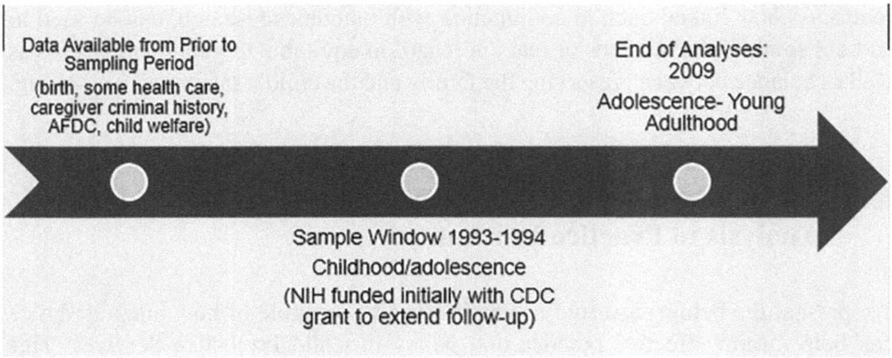

Fig. 6.1 Integrated data example

Baseline Group	Later CAN	Later AFDC	Remain "Original Status"
Poverty Only (n=5107)	1695	--	3412 (67%)
CA/N Only (n=2234)	--	383	1851 (83%)
CA/N & AFDC (n=5073)	--	--	5073 (100%)

Fig. 6.2 Example of changes in sample conditions over time

based on analyses done in response to an agency director's interest in the idea of an iceberg theory regarding child maltreatment report recidivism (Jonson-Reid et al., 2003). There was also a great deal of communication between agencies about the definitions and best uses of variables by the agency who entered the data (Jonson-Reid & Drake, 2008).

Indeed, there were many kinds of data cleaning and enhancement done over the course of this project. When you think about integrated data over time, you have to account for the fact that people may move in and out of original sample groups as new needs or systems contacts might occur. For example, when we selected our sample, we had "clean" groups which were either present in the AFDC/TANF data but not in the CPS data ("Poverty Only") or present in the CPS data but not in the AFDC/TANF data ("Child Abuse and Neglect (CA/N) Only") or present in both ("CA/N & AFDC"). These groups, however, shifted over time. See the example below (Fig. 6.2). This kind of mobility over time is important to consider in administrative data and poses methodological and interpretive challenges. Within the poverty-only group, some children were later reported for alleged maltreatment. Some of the comparison group of children reported for maltreatment lacked histories of income maintenance; later we had families move on to income maintenance. These factors may relate to the outcomes of interest. For example, children who continued to end up back in the child protection system for new reports had far worse outcomes across a number of domains throughout adolescence and young adulthood (Jonson-Reid et al., 2012).

Back to Our Story One area of substantive focus of our study had to do with the relationship between child welfare involvement and special education. What little research that existed suggested an overlap between the populations (Sullivan & Knutson, 2000). Our agency partners were interested in learning more about this. There was a general assumption that children entered school, were identified as having special needs, and then at some point, educators reported a concern to child protective services. In other words, the idea was that child maltreatment concerns first became visible because of attending school. This made intuitive sense as concerns about maltreatment required that a concerned person see a child, most children attend school starting at about age five, and educators are the most common professional reporters (US DHHS, 2023). Because of this, it was expected that the order of agency contacts for a child with a known disability was first school and then child protection.

What we found ended up being the exact opposite. Our analyses indicated that the majority of children eligible for special education had a first report in early childhood, well before they contacted the public school system. There was usually a large and protracted gap between their first contact with the child welfare system and any later contact they may have had from special education systems (Jonson-Reid et al., 2004). Science on child development tells us that though we might not be able to offset every possible disability, earlier identification and intervention was and continues to be associated with improved outcomes (Macy et al., 2014; Singh & Anekar, 2018). Given the science on the importance of early detection and intervention and the gap our data showed, there was an obvious opportunity to provide preventative services. What if we could identify these families earlier as a result of their initial child welfare contact and try to get them special education services sooner?

Working with integrated data is not, and should never be considered, a non-field exercise. Our team took the information we had developed back to the partner agencies and asked about their view of the gap that we saw in the data. After all, maybe families were getting services, but we just lacked the records about those services. We did have at least three models of home visiting in the region at the time, and we did not have their data. So, we conducted focus groups, held meetings, and talked to lead staff to try to understand whether families were being connected to early childhood services and, if not, why not (Stahlschmidt et al., 2018). We considered linking the data from the largest local home visiting group to child protection data, but they only collected de-identified summaries at the central office. We were able to look at summaries of their risk factors, however, and it did not look like the families that child welfare was seeing were in their population. Our best guess was that approximately 10% of these child welfare families with young children at risk for developmental disability were being connected with appropriate support services. There were a lot of reasons for this phenomenon—largely related to the training of child welfare staff, the understanding of child protection by early childhood agencies, and the typical means of recruiting families into early childhood home-visiting services. It became obvious that despite services being available, the knowledge and processes to create an effective bridge from the population encountered by child protection to the developmental services that they needed simply did not exist. After much discussion, we developed a "warm handshake" model to try to connect the population of very young children who were encountered by CPS to early childhood services. After a long search for funding, we were able to complete a small effectiveness study of the Early Childhood Connections project (Jonson-Reid et al., 2018).

But that is, of course, not the end of the story in regard to linked data. Up until now, we have been discussing the value of linked administrative data in identifying a need and crafting a possible solution. We intentionally linked administrative data to the data collected in the field, which both allowed for us to measure an agency-relevant measure of repeat reports to child protection and helped offset concerns with attrition. While most early childhood models are designed to be of long term (1–2 years or more), most families drop out of such programs much earlier (Chiang et al., 2018). One of the other valuable aspects of integrating administrative data into intervention studies is that, while attrition often impacts the ability to continue to collect interview data, it does not impact the ability to measure outcomes available in administrative records.

Administrative records can also help reveal unforeseen characteristics of subjects enrolled in a study. For example, our project was designed to intervene at the earliest contact with CPS, but after data were linked, we discovered a significant portion of the families involved in the study were in fact not first-time CPS contacts. Some caregivers had contact years prior with children that had been removed from care, but this information held in state records was not readily available to the county CPS investigator that was providing data for our project. It turned out that controlling for prior history was extremely important in understanding whether a new intervention, in this case home visiting, decreased recurrent reporting (Jonson-Reid et al., 2018).

In this example, administrative data was essential at all stages of the research process. A large integrated data set helped us to understand a potential problem and window of opportunity for intervention. The process of collaboration to obtain and report on the data provided a ready set of contacts to try to develop a solution. Then the linkage of administrative data to the field data collected to test the solution helped us understand why particular subgroups did or did not benefit from the intervention.

6.3 An Overview of Uses of Linked Administrative Data in Child Welfare Research

The core focus of this chapter is to describe how linked administrative data can be used to find a better balance between the need for, amount of, and timing of services provided and the effectiveness and cost of services. Our case example illustrated one way such data can be used to identify a need for and potential better timing of services. However, there are many other ways in which linked administrative data can be used in furtherance of these goals. At an epidemiological level, these data can be used to address issues such as identifying populations at risk to target prevention (e.g., Putnam-Hornstein & Needell, 2011), understanding trajectories of individuals or families as they encounter various service systems (e.g., Marshall et al., 2011), or amplifying connections (or lack of connections) between existing systems (e.g., Jonson-Reid et al., 2004). In terms of framing and implementing interventions or policies, linked administrative data have other advantages. Many children's service systems are dependent on other systems to meet client needs, but rarely have access to information about what services other organizations may be providing. This is not only a problem for effective case management (Dehart & Shapiro, 2017; Jonson-Reid et al., 2017) but can also lead to funders and policy makers making inappropriate decisions about what is or is not working (Jonson-Reid, 2011). Administrative data research properly conducted in partnership with agencies can help identify service trajectories across systems and existing gaps at a more rapid pace and usually much lower cost than traditional methods (Dehart & Shapiro, 2017). When evaluating services, linked administrative data can provide a rich empirical context to better understand which services are effective for which populations under which circumstances. Large linked data systems also have the potential to allow for rich exploration of subpopulations that are rarely present in sufficient numbers in even the largest survey datasets (Connelly et al., 2016). Below, we deal with several of these issues in turn.

6.3.1 Improving Our Ability to Understand Client Service Use

Case management programs have the stated goal of helping people by connecting them to needed services. In order to do this efficiently, the case worker needs to answer a range of questions: (1) What is the client's past history of accessing these services? (2) Do the services even exist for the client to access? (3) Is the client using or continuing to use these services? Child protective services (and many other systems as well) are largely based on such a case management approach. Even a case worker highly skilled in case management—no matter how well they bond with the family or how good the assessment of need is—will have outcomes that are largely dependent on external resources (e.g., services from other agencies that are direct providers). Not only do agencies not always do a good job of describing what services are provided, it is also often unclear what clients are able to successfully access once they are referred (Jonson-Reid et al., 2017). Families may also have prior histories with services within the same or even prior generations that can impact their willingness to access or ability to benefit from services (Jonson-Reid et al., 2018; Marshall et al., 2011). Not having access to these data can lead to conclusions that child protective services are ineffective, when the actual problem may be lack of availability of services beyond CPS that are critical to address risks leading to maltreatment onset, recurrence, and/or family reunification (Choi & Ryan, 2007; Jonson-Reid, 2011). Integrated data can help researchers understand what happens after a referral and what combination of services and length of services across agencies may result in the best outcome.

6.3.2 Improving Understanding of Rare Events and Subpopulations of Interest

While large-scale survey studies like the National Study of Child and Adolescent Well-being exist, they often fail to capture sufficient numbers of marginalized groups (e.g., minoritized ethnic/racial groups or immigrant populations) to understand variations in system contact or services. For example, in most surveys, many distinct ethnic groups are lumped together into broad categories despite significant differences in culture and nativity. By using linked administrative data, one can capture both smaller subgroups and a measure of US-born compared to foreign-born populations that can illuminate the need for variation in practice—sometimes having access to the full child welfare population (Finno-Velasquez et al., 2017; Johnson-Motoyama et al., 2015).

Similarly rare but important events like child fatality are often unable to be studied in surveys. While system data may be useful for studying deaths that are reported as child maltreatment fatalities (Farrell et al., 2017), they cannot shed light on other forms of preventable death among children with CPS contact. Integrated administrative data can be used to identify such deaths and also what systems are best poised to prevent such outcomes (Jonson-Reid et al., 2017; Wall-Wieler et al., 2018).

6.3.3 Adding Contextual, Explanatory, or Control Variables

One of the basic necessities in research is achieving what researchers call "Content Validity." This is the notion that if you are trying to understand something, you are unlikely to understand it well (either conceptually or in a statistical model) if you don't include all the necessary ideas or measures. For example, apparent disproportionalities among child protective services among different racial or ethnic groups generally change radically when factors such as income or prior risk are included in statistical models (Putnam-Hornstein et al., 2013). In simple terms, if you don't have all the pieces, the puzzle may not come together. Linked data provide an effective means of better understanding whether we are over- or underserving a given population within or across agencies. For example, by linking to law enforcement, health, education, income support, and other data sets, we can examine things like whether child protection reports appear to be results of over-surveilling the marginalized groups or useful tools for identifying families in need of support (e.g., Jonson-Reid et al., 2009). Sometimes even additional single sources of data, such as birth records, can provide a tremendous amount of information and can substantially inform our ability to understand risk (Putnam-Hornstein & Needell, 2011). A very overlooked source of data is simple community contextual factors, which are publicly available and can be found in the Census. Knowing a person's address can provide a simple way to characterize families across a variety of domains—because community SES is a powerful predictor of a range of risk factors and outcomes (Kim & Drake, 2018; Pelton, 2015). Geographic variables can also help researchers understand service-access issues in a region that may in turn impact case outcomes (Friesthler, 2013).

6.3.4 Characterizing Treatment as Usual

In the context of child protective services, "usual care" can be considered a process that begins with the screening-in of a child maltreatment report, through the assessment/investigation and the final decision/outcome of that report (Jonson-Reid et al., 2017). There is significant variability in "standard practice" within child welfare systems at the state and county levels. More specifically, child welfare systems can vary in how they define and respond to allegations of child maltreatment, as well as in the support they offer to children and families (Jonson-Reid & Chiang, 2019). For example, South Dakota screens out 84.7% of their CPS referrals, while Mississippi screens out only 24% (US DHHS, 2023). This variability often makes it difficult to understand what "usual care" in child protection looks like. The lack of a common definition of "usual care" poses several challenges and "can lead to under or over-estimation of effects, poor understanding of program coverage by policy makers, and inadequate ability to track changes that result from reform initiatives" (Jonson-Reid et al., 2017, p. 222).

From a policy perspective, it is obviously necessary to understand current baseline practice. Evidence-based practice and policy cannot exist without a firm empirical basis (Landsverk et al., 2005). Without a strong understanding of what is happening in an agency, an agency cannot be accountable for current practice and cannot plan intelligently to improve practice. While child protective agencies are federally required to monitor specific outcomes as a function of mandated SACWIS (Statewide Automated Child Welfare Information System) structure and compliance with the federal CFSR (Child and Family Service Review) audits, these benchmarks do not capture the totality of child protective services in a useful way and are not alone sufficient to inform agency efforts to improve policy or practice.

From a research perspective, ethical considerations generally preclude the possibility of comparing any innovative service to a "no service" condition. Because of this, child welfare policy and practice innovations are typically compared to a "treatment as usual" condition (Jonson-Reid & Chiang, 2019). In simple terms, the effectiveness of any intervention is therefore the degree to which it improves upon (or fails to improve upon) treatment as usual. Integrated administrative data offer a means of observing usual care across systems as a baseline for understanding whether an additional intervention has added value. This is essential as usual care can vary across time as caseloads and resources change.

6.3.5 *Identifying Policy Relevant Concerns: The Case of Substantiation in CPS*

Fields of practice often use ideas or terms which fundamentally define and configure what they do. In Child Protective Services, one of the key labels used by many states with significant policy import is "substantiation." Understanding this term is complicated by the various ways it is used in different states. Some states use the term "substantiated" while others use terms like "founded" or "indicated." Some states have "three tier" systems, in which cases can be either unfounded or substantiated or may classed as an intermediate "substantiation-lite" type strata. Still other states use "alternative response" or "differential response" systems in which cases are triaged into "investigation" and "assessment" tracks, and the (generally less severe) "assessment" cases may not get a determination of substantiation at all. In any case, there has been a general long-standing assumption that "substantiated" cases are analogous to a "guilty" verdict and that "unsubstantiated" cases mean "nothing happened" (Drake, 1996). This assumption drives any number of state policies, ranging from gating services to assignment to a central registry that can be used for employment checks. Recently, however, it has become clear that the "substantiated"/ "unsubstantiated" dichotomy is not very useful. It simply does not predict future risk to the child. The reason we now know this is because of linked administrative data.

Labels can matter. If one looks up the most recent national Child Maltreatment report (US DHHS, 2023), one will see that descriptive information on children reported is limited to "victims." This is an unfortunate word that is tied to the legal standard for substantiation (which again varies by state). Substantiation is a legal term meaning that there is both sufficient harm and evidence tied to an alleged perpetrator to allow the possibility of mandatory services or family court involvement (Drake, 1996). Why do we know this label is not synonymous with a child having experienced some level of maltreatment? With few exceptions, studies linking report disposition to later recurrence controlling for poverty and other factors (e.g., Drake et al., 2003; Putnam-Hornstein et al., 2015) and studies linking report disposition to other outcomes like behavioral health or school outcomes (Jaffee & Gallop, 2007; Jonson-Reid et al., 2004; Kohl et al., 2009; Leiter et al., 1994) show very little difference in later outcomes that can be attributed to substantiation status.

These findings call into question a range of policy, research, and practice issues. For example, some states only provide services to substantiated cases. Given that substantiated and unsubstantiated cases are at similar risk of future maltreatment, and that both could therefore benefit from service provision, this policy makes little sense. From a social work or public health perspective, it is important to have assessments that help match services to meet the needs of families (Choi & Ryan, 2007; Drake & Jonson-Reid, 2000). While there may be a need to have an indicator of substantiation for moving to involvement with the court, it is not appropriate as a sole indicator for whether to provide services to a family. To a policy maker, equating substantiation with victimization radically changes one's perception of the size and costs associated with failing to prevent or intervene. Reliance on substantiation leads to a lifetime prevalence of 1 in 8 children, but if we include unsubstantiated cases, this rises to 1 in 3 (Kim et al., 2018; Wildeman et al., 2014). Likewise, estimates of downstream costs more than double between those limited to annual numbers of substantiated compared to all screened-in reports (Fang et al., 2012; Peterson et al., 2018). This can have a serious impact on advocacy for prevention and intervention funding. Finally, many of the large, longer-term studies informing our understanding of outcomes of maltreatment have unfortunately used substantiated reports to indicate the presence or absence of maltreatment in their samples (e.g., Thornberry et al., 2010; Widom, 2019). It is not known what the impact of having unsubstantiated cases of maltreatment in the comparison condition may have had on our understanding of the magnitude of impact of maltreatment on developmental trajectories.

6.3.6 Effectiveness Studies

Administrative data can be extremely valuable in understanding service effectiveness and cost. Even the simple use of a state's unlinked CPS reports can meaningfully increase understanding of differences in maltreatment reporting over time related to policy changes, or across regions or by client characteristics. One recent example involves an evaluation of differential response programs by combining three types of data. Differential response is an approach to "tracking" cases into either an "investigations" track or a "assessment" track. Fluke et al. (2019) started with counts of CPS reports and substantiated reports, which are aggregated in the National Child Abuse and Neglect Data System "Child File" data. They then linked these data to census data and also lists describing which counties had or had not implemented differential response programs. Using this simple approach to linking data at the county level, the team was able to build statistical models that provided compelling preliminary evidence of a lower rate of recurrence in counties that had implemented differential response programs.

Another example of using integrated data to understand service effectiveness can be found in Prinz's well-known studies on the effectiveness of Triple P in South Carolina (Prinz et al., 2009, 2016). This example illustrates one of the simplest examples of integrated data—using official reports of maltreatment as an outcome variable in an experimental design. In this case, counties were randomized to treatment (Triple P) or control conditions (Service as usual). Differences in rates of reported maltreatment were used as the dependent measure to determine whether Triple P was effective in reducing maltreatment. Another randomized study using linked administrative data (Cancian et al., 2013) appended state child maltreatment data to data from a study in which some families received more money than other families on a random basis. This study was able to show that increased financial resources seemed to reduce rates of reported maltreatment, a finding which has been replicated by other studies combining administrative and other forms of data (McLaughlin, 2017).

6.3.7 Understanding Possible Bias in Services

While trying to find families in need or protect children reported, child welfare workers are often put in the middle of arguments over how much intrusion is too much and whether it is equitable across class, race, or other categories. For example, Jonson-Reid et al. (2009) linked a wide range of different data sources to answer the following question: "Are poor children who come to the attention of CPS different from poor children not identified by CPS?" The rationale for this study was that if children were simply being referred to CPS because they were poor, and not for reasons of risk, then steps should be taken to reduce over-reporting of poor children to CPS. If this is true, poor children who were or were not identified by CPS should not look very different in terms of other later outcomes known to be associated with maltreatment like juvenile justice, mental health, disability, or hospital treatment for fractures, etc. Because many of these areas were not subject to reporting bias (e.g., ER admissions, hospital births, etc.), findings would allow for between-group comparisons of risk. The study found that CPS-engaged families showed higher risks across all domains, suggesting that the poor children identified by CPS were qualitatively and broadly at higher risk across domains than children not identified by the CPS system. A very different approach was taken by Kim, Drake, and Jonson-Reid (2018), who linked census data to Child File data to answer an associated question: "Do poorer counties show higher rates of reports from mandated sources?" The logic underlying this question was that if poor people were more likely to be reported due to their increased visibility to mandated reporters, then the proportion of reports from mandated sources would have to be higher in poorer areas. This study used national data and had the surprising finding that the proportion of reports from mandated sources actually decreases with increasing poverty, a result inconsistent with the theory of class-based bias.

Similarly, linked data is now being used to explore racial bias in reporting and child protection decision-making at the population level as well as within known risk groups. Racial disproportionality is a significant policy and ethical concern stemming from comparing rates of system involvement to prevalence in the population. As sociodemographic and other risks are controlled by linking birth or income or other records, however, there appears to be increasing evidence that sometimes ethnic minority children are under-reported and underserved (Putnam-Hornstein et al., 2013, 2016). The disparity in risk, particularly poverty, driving disproportionality has been supported by data from surveys, and other sources indicate that when disproportionality exists, it is largely or completely explained by disparities in risk rather than by differential treatment (Font et al., 2012; Drake et al., 2011).

In summary, linked administrative data is extremely flexible, and not simply a medium for descriptive or epidemiological work.

6.3.8 *Possible Perils*

What about the dark side of integrated data? Data misuse or inaccuracies, security breaches, results we don't want to see—these are all issues that are raised when people consider constructing and using integrated data sets. First, all data, whether from in-person surveys, intervention studies, or administrative data, have flaws. Integrated data approaches do not replace other types of data, but they are often the most effective and efficient means of answering policy-relevant questions quickly and accurately (Connelly et al., 2016). No form of data or research method is perfect, but the degree to which radically different data and approaches are used supports our ability to triangulate. Ultimately, the more we triangulate between data sources and studies, the closer we get to the truth.

6.3.8.1 Data Misuse

There are often concerns that data somehow produce surveillance or other system bias just by connecting them. Actually, the opposite is arguably true. If one knows what systems and communities a family is involved with at a more comprehensive level, it is actually easier to make decisions more transparent and look for places where such problems might be evident rather than just guess (Dehart & Shapiro, 2017). Linked data can serve as a source of evidence to evaluate a range of critical ethical and value-related questions associated with child welfare practice and policy (Jonson-Reid & Drake, 2008).

6.3.8.2 Data We Do Not Want to See

Administrative data can highlight things we do not want to see as well as those researchers do, but can produce anxiety among stakeholders (Donaldson et al., 2002). While avoiding sharing services data that may show less than desirable outcomes may seem like a good strategy, it typically comes out in some way in the long run. At least having the information and knowing that a program is not working along the way allows an agency time to course correct and allows them time to manage concerns proactively. In other words, it gives an agency the opportunity to think of more preventative fixes (Dehart & Shapiro, 2017).

6.3.8.3 Security Breaches

Linked data also are the subject of a range of ethical concerns about the creation and use of such data. One common concern revolves around the possibility of security breaches. Linked data is not inherently less secure than unlinked data. All service providers are keeping data on the families they work with. The fact that it is shared in archives that can produce de-identified data for analysis does not make it more at risk than when the data was unlinked and stored separately before. Additionally, university partners are often much more regulated and surveilled by many other organizational regulatory forces that can impose significant career penalties if we somehow have a breach in security. Finally, there is little financial incentive to attempt to hack a system following families in services. Unlike purchases on Amazon or credit reports or banking systems, there is little to be gained from a criminal perspective in trying to access records on a largely very low-income population. As pointed out by Culhane and colleagues (2018), many integrated data systems have been operating for decades and already include a significant portion of the population.

6.3.8.4 Missing Information

Another concern around data is on missing information (Connelly et al., 2016). Data can be missing due to poor recording practices but may also be impacted by policy shifts. States and agencies may keep data for varying periods of time. For example, in 2003, they changed the expungement protocols in the state of Missouri. One of the reasons why our child welfare workers did not always know whether families in the case study example were first-time reports was because unsubstantiated cases were sometimes expunged. Another reason for this lack of information was that families may have moved out of the service area but such information may not be known or recorded in administrative records. Parrish et al. (2017) estimated the level of bias associated with out-of-state emigration on prevalence and model parameters to be generally modest, but it is an important consideration. Here again, there is nothing particularly special about administrative data. Data can also be missing from intervention studies due to attrition, from surveys due to non-response, or from interviews due to refusal to answer a question. Sometimes linked data can be a useful tool to address what may be missing in one system but not in another (Jonson-Reid & Drake, 2008). For example, perhaps education is not recorded in child welfare but is recorded in income maintenance, so by linking the two you no longer have missing information. This triangulation is a powerful tool for leveraging the complementary strengths of different data systems. Like any data set, you have to know how the data were collected and when data are missing and why.

6.3.8.5 The Politics and Partnerships of Making It Work

Gaining access to data to be linked can be challenging. Trust-building and discussions prior to creating linked data can help ally some of these fears and improve the likelihood that the information will lead to positive change (Davern et al., 2017). One builds this trust by talking through both the process of sharing and security, but also the use and dissemination agreements (Culhane et al., 2018). Some agencies may be concerned that they cannot share data due to lack of individual consent, but this is actually a myth in most cases. As Culhane et al. (2018) discuss, most agencies can share data to improve services or inform accountability without consent of the party receiving the service. This is consistent with the practice of sharing back information (positive and negative) to the agency stakeholders so that information can be used to improve services, not just inform research (Dehart & Shapiro, 2017; Jonson-Reid & Drake, 2008).

While some characterize administrative data as "found" (Connelly et al., 2016), ongoing partnerships with agency stakeholders can also be leveraged to alter what is collected by data systems. In the past, many large organizations were forced to pay for expensive third-party data systems. If you changed a question, that was a huge cost burden. This is changing. There are much more user-friendly ways of doing this now that provide an opportunity to improve the quality and content of data we collect and store electronically. In some cases, there may actually be long-term cost savings if you are adding information that makes it easier to report out required metrics on child well-being (not typically present in a single dataset) for the Child and Family Services Reviews (https://www.cfsrportal.acf.hhs.gov/).

Political environments do change. Findings from the linked data study presented earlier were well received by agencies who posed additional questions in meetings. To answer some of these would require ongoing data collection, and for a while it seemed that everyone was on board. Then our state administration shifted toward increasing attention to risk management and a large-scale effort to privatize child welfare. In addition, when the recession hit, they radically downsized the research and evaluation staff. All of these factors influenced the willingness and capacity to share data. The tide is again changing and this may become a reality in our state again, but it is important to consider when planning so that Memorandum of Understanding and Data Use Agreements exist that can help transcend inevitable shifts in administration.

6.4 Future Directions

Other chapters in this volume attend to details about studies or methods related to integrated or linked administrative data. The call for its use to inform practice and policy has grown exponentially in the last decade. The question is now changing from whether you will create and use linked administrative data to how. There are many more models available now and funded projects underway to enhance and make easier the use of integrated data in child welfare. While there are certainly barriers and pitfalls, the same is true of all research endeavors. We view this change as a positive move toward finding the "Goldilocks Zone" for improving services for children and families.

References

Cancian, M., Yang, M. Y., & Slack, K. S. (2013). The effect of additional child support income on the risk of child maltreatment. *Social Service Review, 87*(3), 417–437.
Chiang, C., Jonson-Reid, M., Kim, H., Drake, B., Kohl, P., Pons, L., Constantino, J., & Auslander, W. (2018). Service engagement and retention in home visitation: Lessons from the early childhood connections program. *Children and Youth Services Review., 88*, 114–127.
Choi, S., & Ryan, J. P. (2007). Co-occurring problems for substance abusing mothers in child welfare: Matching services to improve family reunification. *Children and Youth Services Review, 29*(11), 1395–1410.
Chouldechova, A., Benavides-Prado, D., Fialko, O., & Vaithianathan, R. (2018, January). A case study of algorithm-assisted decision making in child maltreatment hotline screening decisions. In Conference on fairness, accountability and transparency. *Proceedings of Machine Learning Research, 81*, 134–148.
Connelly, R., Playford, C. J., Gayle, V., & Dibben, C. (2016). The role of administrative data in the big data revolution in social science research. *Social Science Research, 59*, 1–12.
Culhane, D., Fantuzzo, J., Hill, M., & Burnett, T. C. (2018). Maximizing the use of integrated data systems: Understanding the challenges and advancing solutions. *The Annals of the American Academy of Political and Social Science, 675*(1), 221–239.
Davern, M. T., Gunn, L., Giles-Corti, B., & David, S. (2017). Best practice principles for community indicator systems and a case study analysis: How community indicators Victoria is creating impact and bridging policy, practice and research. *Social Indicators Research, 131*(2), 567–586.
DeHart, D., & Shapiro, C. (2017). Integrated administrative data & criminal justice research. *American Journal of Criminal Justice, 42*(2), 255–274.
Donaldson, S. I., Gooler, L. E., & Scriven, M. (2002). Strategies for managing evaluation anxiety: Toward a psychology of program evaluation. *American Journal of Evaluation, 23*(3), 261–273.
Drake, B. (1996). Unraveling "unsubstantiated". *Child Maltreatment, 1*(3), 261–271.
Drake, B., & Jonson-Reid, M. (2000). Substantiation, risk assessment and involuntary versus voluntary services. *Child Maltreatment, 5*(3), 227–235.
Drake, B., Jolley, J., Lanier, P., Fluke, J., Barth, R., & Jonson-Reid, M. (2011). Racial bias in child protection? A comparison of competing explanations using national DATA. *Pediatrics, 127*(3). https://doi.org/10.1542/peds.2010-1710. PMID:21300678.
Drake, B., Jonson-Reid, M., Way, I., & Chung, S. (2003). Substantiation and recidivism. *Child maltreatment, 8*(4), 248–260.
Fang, X., Brown, D. S., Florence, C. S., & Mercy, J. A. (2012). The economic burden of child maltreatment in the United States and implications for prevention. *Child Abuse & Neglect, 36*(2), 156–165.
Farrell, C. A., Fleegler, E. W., Monuteaux, M. C., Wilson, C. R., Christian, C. W., & Lee, L. K. (2017). Community poverty and child abuse fatalities in the United States. *Pediatrics, 139*(5), e20161616.
Finno-Velasquez, M., Palmer, L., Prindle, J., Tam, C. C., & Putnam-Hornstein, E. (2017). A birth cohort study of Asian and Pacific islander children reported for abuse or neglect by maternal nativity and ethnic origin. *Child Abuse & Neglect, 72*, 54–65.
Fluke, J. D., Harlaar, N., Brown, B., Heisler, K., Merkel-Holguin, L., & Darnell, A. (2019). Differential response and children re-reported to child protective services: County data from the National Child Abuse and Neglect Data System (NCANDS). *Child Maltreatment, 24*(2), 127–136.
Font, S. A., Berger, L. M., & Slack, K. S. (2012). Examining racial disproportionality in child protective services case decisions. *Children and Youth Services Review, 34*(11), 2188–2200.

Freisthler, B. (2013). Need for and access to supportive services in the child welfare system. *GeoJournal, 78*(3), 429–441.

Gary, S. (2016, February). *What is the Goldilocks Zone and why does it matter in the search for ET?* https://www.abc.net.au/news/science/2016-02-22/goldilocks-zones-habitable-zone-astrobiology-exoplanets/6907836

Jackson, P. (2019). From 'intruders' to 'partners': The evolution of the relationship between the research community and sources of official administrative data. In *Data-driven policy impact evaluation* (pp. 17–26). Springer.

Jaffee, S. R., & Gallop, R. (2007). Social, emotional, and academic competence among children who have had contact with child protective services: Prevalence and stability estimates. *Journal of the American Academy of Child and Adolescent Psychiatry, 46*, 757–765.

Johnson-Motoyama, M., Putnam-Hornstein, E., Dettlaff, A. J., Zhao, K., Finno-Velasquez, M., & Needell, B. (2015). Disparities in reported and substantiated infant maltreatment by maternal Hispanic origin and nativity: A birth cohort study. *Maternal and Child Health Journal, 19*(5), 958–968.

Jonson-Reid, M., & Chiang, C. (2019). Problems in understanding program efficacy in child welfare. In B. Lonne, D. Scott, D. Higgins, & T. Herrenkohl (Eds.), *Visioning public health approaches for protecting children*. Springer.

Jonson-Reid, M., & Drake, B. (2008). Multi-sector longitudinal administrative databases: An indispensable tool for evidence-based policy for maltreated children and their families. *Child Maltreatment, 13*(4), 392–399.

Jonson-Reid, M., Drake, B., Chung, S., & Way, I. (2003). Cross-type recidivism among referrals to a state child welfare agency. *Child Abuse & Neglect, 27*, 899–917.

Jonson-Reid, M., Drake, B., Kim, J., Porterfield, S., & Han, L. (2004). A prospective analysis of the relationship between reported child maltreatment and special education eligibility among poor children. *Child Maltreatment, 9*, 382–394.

Jonson-Reid, M., Drake, B., & Kohl, P. L. (2009). Is the overrepresentation of the poor in child welfare caseloads due to bias or need? *Children and Youth Services Review, 31*(3), 422–427.

Jonson-Reid, M. (2011). Disentangling system contact and services: A key pathway to evidence-based children's policy. *Children and Youth Services Review, 33*(5), 598–604.

Jonson-Reid, M., Kohl, P. L., & Drake, B. (2012). Child and adult outcomes of chronic child maltreatment. *Pediatrics, 129*(5), 839–845.

Jonson-Reid, M., Drake, B., & Kohl, P. L. (2017). Childhood maltreatment, public service system contact, and preventable death in young adulthood. *Violence and Victims, 32*(1), 93–109.

Jonson-Reid, M., Drake, B., Constantino, J., Tandon, M., Pons, L. Kohl, P., Roesch, S., Wideman, E., Dunnigan, A. & Auslander, W. (2018). A randomized trial of home visitation for CPS involved families: The moderating impact of maternal depression and CPS history. *Child Maltreatment*. Advanced online: http://journals.sagepub.com/doi/pdf/10.1177/1077559517751671

Kim, H., & Drake, B. (2018). Child maltreatment risk as a function of poverty and race/ethnicity in the USA. *International Journal of Epidemiology, 47*(3), 780–787.

Kim, H., Drake, B., & Jonson-Reid, M. (2018). An examination of class-based visibility bias in national child maltreatment reporting. *Children and Youth Services Review, 85*, 165–173.

Kohl, P. L., Jonson-Reid, M., & Drake, B. (2009). Time to leave substantiation behind: Findings from a national probability study. *Child maltreatment, 14*(1), 17–26.

Landsverk, J., Barth, R. P., Jones Harden, B., & Yuan, Y.-Y. T. (2005). *Beyond common sense: Child welfare, child well- being, and the evidence for policy reform*. AldineTransaction.

Leiter, J., Myers, K. A., & Zingraff, M. T. (1994). Substantiated and unsubstantiated cases of child maltreatment: Do their consequences differ? *Social Work Research, 18*(2), 67–82.

Macy, M., Marks, K., & Towle, A. (2014). Missed, misused, or mismanaged: Improving early detection systems to optimize child outcomes. *Topics in Early Childhood Special Education, 34*(2), 94–105. https://doi.org/10.1177/0271121414525997

Marshall, J. M., Huang, H., & Ryan, J. P. (2011). Intergenerational families in child welfare: Assessing needs and estimating permanency. *Children and Youth Services Review, 33*(6), 1024–1030.

McFall, M. (2009). *Licensing parents. Family, state and child maltreatment.* Lexington Books.

McLaughlin, M. (2017). Less money, more problems: How changes in disposable income affect child maltreatment. *Child Abuse & Neglect, 67*, 315–321.

Parrish, J. W., Shanahan, M. E., Schnitzer, P. G., Lanier, P., Daniels, J. L., & Marshall, S. W. (2017). Quantifying sources of bias in longitudinal data linkage studies of child abuse and neglect: Measuring impact of outcome specification, linkage error, and partial cohort follow-up. *Injury Epidemiology, 4*(1), 23.

Pelton, L. H. (2015). The continuing role of material factors in child maltreatment and placement. *Child Abuse & Neglect, 41*, 30–39.

Peterson, C., Florence, C., & Klevens, J. (2018). The economic burden of child maltreatment in the United States, 2015. *Child abuse & neglect, 86*, 178–183.

Prinz, R. J., Sanders, M. R., Shapiro, C. J., Whitaker, D. J., & Lutzker, J. R. (2009). Population-based prevention of child maltreatment: The US Triple P system population trial. *Prevention science, 10*, 1–12.

Prinz, R. J., Sanders, M. R., Shapiro, C. J., Whitaker, D. J., & Lutzker, J. R. (2016). Addendum to "Population-based prevention of child maltreatment: The US Triple P system population trial". *Prevention Science, 17*(3), 410–416.

Putnam-Hornstein, E., & Needell, B. (2011). Predictors of child protective service contact between birth and age five: An examination of California's 2002 birth cohort. *Children and Youth Services Review, 33*(8), 1337–1344.

Putnam-Hornstein, E., Needell, B., King, B., & Johnson-Motoyama, M. (2013). Racial and ethnic disparities: A population-based examination of risk factors for involvement with child protective services. *Child Abuse & Neglect, 37*(1), 33–46.

Putnam-Hornstein, E., Simon, J. D., Eastman, A. L., & Magruder, J. (2015). Risk of re-reporting among infants who remain at home following alleged maltreatment. *Child Maltreatment, 20*(2), 92–103.

Putnam-Hornstein, E., Prindle, J. J., & Leventhal, J. M. (2016). Prenatal substance exposure and reporting of child maltreatment by race and ethnicity. *Pediatrics, 138*(3), e20161273.

Singh, P., & Anekar, U. (2018). The importance of early identification and intervention for children with developmental delays. *Indian Journal of Positive Psychology, 9*(2), 233–237. Retrieved from http://search.ebscohost.com/login.aspx?direct=true&db=a9h&AN=133608408&site=ehost-live&scope=site

Stahlschmidt, M., Jonson-Reid, M., Pons, L., Constantino, J., Kohl, P., Drake, B., & Auslander, W. (2018). Trying to bridge the worlds of home visitation and child welfare: Lessons learned. *Evaluation and Program Planning, 66*, 133–140.

Sullivan, P. M., & Knutson, J. F. (2000). Maltreatment and disabilities: A population-based epidemiological study. *Child Abuse & Neglect, 24*(10), 1257–1273.

Thornberry, T. P., Henry, K. L., Ireland, T. O., & Smith, C. A. (2010). The causal impact of childhood-limited maltreatment and adolescent maltreatment on early adult adjustment. *Journal of Adolescent Health, 46*(4), 359–365.

US DHHS (2023) Child Maltreatment 2021. https://www.acf.hhs.gov/cb/report/child-maltreatment-2021

Wall-Wieler, E., Vinnerljung, B., Liu, C., Roos, L. L., & Hjern, A. (2018). Avoidable mortality among parents whose children were placed in care in Sweden: A population-based study. *Journal of Epidemiology and Community Health, 72*(12), 1091–1098.

Widom, C. S. (2017). Long-term impact of childhood abuse and neglect on crime and violence. *Clinical Psychology: Science and Practice, 24*(2), 186–202.

Widom, C. S. (2019). Are retrospective self-reports accurate representations or existential recollections?. *JAMA psychiatry, 76*(6), 567–568.
Wildeman, C., Emanuel, N., Leventhal, J. M., Putnam-Hornstein, E., Waldfogel, J., & Lee, H. (2014). The prevalence of confirmed maltreatment among US children, 2004 to 2011. *JAMA pediatrics, 168*(8), 706–713.
Willems, J. C. M. (2019). Children as Subjects of Rights: Three Waves of Emancipation–From past and present cruelty to future creativity? *Net Journal of Social Sciences, 7*, 28–43.
Zeanah, C. H., & Humphreys, K. L. (2018). Child abuse and neglect. *Journal of the American Academy of Child & Adolescent Psychiatry, 57*(9), 637–644.

Chapter 7
Integrating Child Welfare and Medicaid Data to Identify and Predict Superutilization of Services for Youth in Foster Care

Elizabeth Weigensberg

This chapter summarizes findings from the *Superutilization of child welfare, Medicaid, and other services* study, which is referred to as the superutilization study (Weigensberg et al., 2018). This project involved linkage of child welfare and Medicaid data to study high levels of service use or what we refer to as "superutilization" of services among children in foster care. The chapter addresses the following objectives: (1) provide insights about the potential for linking Medicaid and child welfare data, (2) share innovative ways to assess high levels of service utilization across multiple services systems, (3) explain how linked services data can be used to predict key outcomes for children in foster care, and (4) consider how linked Medicaid and child welfare data can inform targeted interventions for children in foster care.

The study is a good example of the potential of linking child welfare data with Medicaid and other health data. One of the opportunities with this study was to look not only at outcomes for children once they leave foster care, but to better understand what services children are receiving while in the foster care system. The rich set of services data allowed us to examine what high service utilization looked like for children in foster care and to explore the use of predictive analytics, allowing us to consider how this information can help target services and interventions for children and families.

For this project, Mathematica collaborated with Casey Family Programs along with the child welfare and Medicaid agencies in Tennessee and Florida. Also, because Florida has a privatized child welfare system, we partnered with the local community-based care provider Eckerd Kids, who served the three-county area of Hillsborough, Pasco, and Pinellas counties, which was the scope of the Florida site for the study. We had a great team of collaborators and partners working on this

E. Weigensberg (✉)
Mathematica, Chicago, IL, USA
e-mail: eweigensberg@mathematica-mpr.com

C. M. Connell, D. M. Crowley (eds.), *Strengthening Child Safety and Well-Being Through Integrated Data Solutions*, Child Maltreatment Solutions Network, https://doi.org/10.1007/978-3-031-36608-6_7

project. Our objective was to ensure the study results would be meaningful and useful for the site partners as well as the broader field.

Early in the project, we produced an issue brief (Maher et al., 2016) outlining the importance of understanding cross-system service utilization across child welfare and health systems. Such a perspective is especially important given children in foster care are eligible for Medicaid to pay for their health care services, yet data sharing across Medicaid and child welfare agencies is rare (Maher et al., 2016). There is a need for information about how child welfare agencies, Medicaid agencies, and related services come together to support the needs of children and families. Linked data is needed to consider interventions at key points that can help prevent children and families from becoming high users of services.

Our study linked child welfare, Medicaid, and other behavioral health data to better understand and predict superutilization of child welfare and other services. Note that superutilization or high levels of service use is not necessarily negative. For those children and families that need it, high levels of service use may be very appropriate to help meet those needs. This study did not place a value judgment on the concept of superutilization. This was an exploratory study of what service use looks like, especially for those who are at higher levels of service usage.

7.1 Study Overview

The study research questions included the following: (1) What is superutilization of child welfare and other services? (2) What are the types of superutilization? (3) What predicts high levels of placement instability at the time of entry into foster care?

We wanted to leverage insights from linked data to understand what service usage looks like for children in foster care. At the start of the project, we hypothesized that the children who use high levels of child welfare services would be the same children that use high levels of physical or behavioral health services. We wanted to understand whether these were the same children with high levels of service use across all systems or if there were different types of superutilization. We also wanted to explore the possibility of using these data to predict superutilization or, more specifically, one of the types of superutilization that we identified. We focused the predictive analysis on one particular type of superutilization that the state partners were most interested in at the time, which was placement instability, meaning those children with a high number of changes in foster care placements.

The data for the study came from several site partners in Tennessee and Florida. In particular, for Tennessee, we linked child welfare data from the Tennessee Department of Children Services and Medicaid claims data from TennCare. For Florida, we linked several sources of data, specifically child welfare data as well as substance abuse and mental health program services data from the Florida

Department of Children and Families, Medicaid claims data from the Agency of Health Care Administration, and child welfare community-based care purchased services from Eckerd Kids. In Florida, we included additional data from substance use and mental health services that are supported by state funds, which supplement what services are available from Medicaid. The department that oversees these programs was eager to link these services data with child welfare data, since the services are located within the same state agency as child welfare, but the department never had the opportunity to bring these data together before this study. Also, in Florida, we received child welfare data from the state agency as well as Eckert Kids, which is the community-based contracted service provider at the local level for the study counties of Hillsborough, Pinellas, and Pasco Counties. More information about the data sources and data linking can be found in the full final report for the study (Weigensberg et al., 2018).

To answer the research questions, we conducted two phases of analyses. In the first phase, we performed descriptive analysis and latent class analysis. We described the characteristics of high service use and identified types, or latent classes, of superutilization. We also described the characteristics related to each type of superutilization we identified. In the second phase of the analysis, we conducted predictive analysis to assess what factors are predictive of superutilization, based on information known at the time of entry into foster care. Given what could be feasibly done within the resources available for the project, we conducted predictive analysis on one type of superutilization, which was identified by the site partners as those children with a high number of foster care placements, which we refer to as placement instability.

7.1.1 Study Sample

For each of the types of analysis, we had to refine the study sample to maximize the use of services data available for the study. This was also necessary given the need to structure the data appropriately for each type of analysis. Overall, we had large sample sizes for both of the study sites. For the Tennessee site, the study sample for the descriptive and latent class analysis included 21,672 children who entered out-of-home custody in child welfare between July 1, 2011, and December 31, 2015. The predictive analysis sample for Tennessee used a sample of 12,056 children who entered out-of-home custody between July 1, 2012, and January 1, 2014. For the study site that included Hillsborough, Pinellas, and Pasco Counties in Florida, which we will refer to as the Florida sample, the descriptive and latent class analysis sample included 6695 children who entered out-of-home custody in child welfare between September 1, 2013, and December 31, 2015. The predictive analysis sample for the Florida site was 8290 children who entered out-of-home custody between January 1, 2012, and January 1, 2014.

7.1.2 Service Use

One of the first things we examined was what services children were receiving while in foster care. The field often analyzes data on foster care placements, but this study had rich services data that helped us understand what services children were receiving in addition to their foster care placements. We analyzed child welfare services such as transportation benefits, clothing assistance, and other kinds of supportive services.

In Tennessee, about 84% of the children in the study sample were receiving child welfare services in addition to foster care placement and case management. About 85% of the Tennessee sample received Medicaid services, where the greatest percentages among those receiving services were for outpatient services (85%) and emergency services (70%).

In the Florida sample, the findings were different, as only about 20% of children received child welfare services, aside from foster care placement and case management services, from the community-based provider. This lower percentage of children receiving services in Florida was confirmed by study contacts in the site, who noted that sometimes these data may not be well captured at the community-based provider level. They also acknowledged that the percentages of children receiving supplemental child welfare support services were generally lower since the community-based providers are not provided much funding to pay for these extra services. In addition, in the Florida sample, over 90% received Medicaid services and about 16% received state-funded mental health or substance abuse services. A more complete depiction of the service use findings is presented in the final report for the study (Weigensberg et al., 2018).

7.1.3 Measuring Superutilization of Services

In order to understand the types of service superutilization, we first had to develop a measure of superutilization. Many economists recommend a measure based on costs; however, given the rich services data we had available for this study, we wanted to develop a more comprehensive measure that would be more meaningful to inform practice beyond the expense of services.

As depicted in Fig. 7.1, we used four dimensions of service utilization that we could measure with the study data, including (1) service frequency to understand how often children received services, (2) duration to see how long they received services, (3) intensity to some extent to identify the percentage of their time in more restrictive settings, such as residential or congregate care, and (4) cost of services. Although not all dimensions were able to be measured for each service type, we used what data we had available to understand what service utilization looks like among these many different dimensions, creating 12 measures of service utilization.

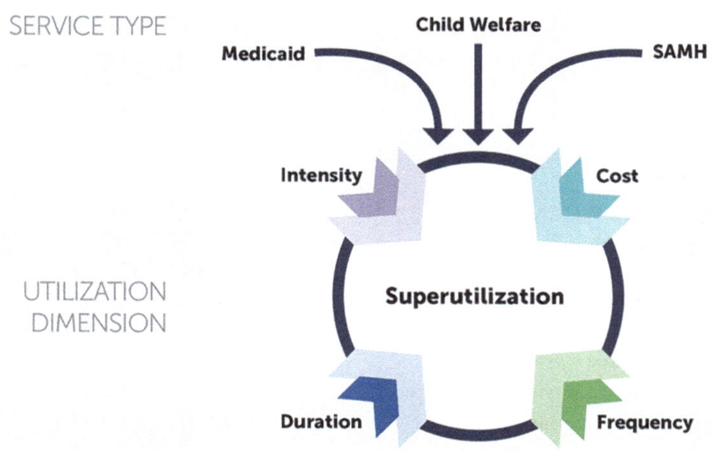

Fig. 7.1 Dimensions of superutilization of services (Note: SAMH refers to the Substance Abuse and Mental Health data in Florida)

Next, we needed to identify a threshold of what is considered "super" in regard to superutilization of services. We used a 90th percentile threshold, so when looking at the distribution of service use for any measure, superutilization was operationalized to reflect those children at the 90th percentile or higher. We also made some adjustments to account for the influence of age and time of exposure to the study window when looking at levels of service utilization. We used this measure to identify superutilization for each of our measures that reflect the multiple domains of service utilization. When looking at the sample of children in the study, over half (56% for the Tennessee sample and 55% for the Florida sample) were identified as passing the superutilization threshold (at the 90th percentile or higher for service use) on any of the measures of superutilization, as shown in Table 7.1.

These findings show us that superutilization does not just occur among a small percentage of the sample; rather, there is a large percentage of children in foster care that can be considered receiving superutilization of services, emphasizing the need to understand the different types of service superutilization.

7.1.4 *Types of Superutilization*

We then used latent class analysis to group children into distinct groups or "classes" based on the observed superutilization measures. The full report (Weigensberg et al., 2018) provides more technical details about our methods, but we identified seven different types of classes or groups of superutilization in the Tennessee sample (Table 7.2) and eight groups of superutilization for the Florida sample (Table 7.3).

Table 7.1 Children in the study sample identified as experiencing superutilization

Measure of superutilization	TN study sample $n = 21{,}672$	FL study sample $n = 6695$
	Number (percent)	Number (percent)
Total number of episodes	3190 (14.7%)	894 (13.4%)
Total number of placement moves	3387 (15.6%)	1078 (16.1%)
Total placement cost per year	2552 (11.8%)	NA
Total episode length of stay	2722 (12.6%)	740 (11.1%)
Average proportion of time in congregate care/residential/group home	1827 (8.4%)	609 (9.1%)
Child welfare/CBC purchased services per year	2432 (11.2%)	601 (9.0%)
Child welfare/CBC purchased services cost per year	1789 (8.3%)	567 (2.6%)
Medicaid inpatient services per year	1257 (5.8%)	380 (5.7%)
Medicaid outpatient services per year	2261 (10.4%)	762 (5.7%)
Medicaid emergency services per year	2213 (10.2%)	380 (5.7%)
Mental health services per year	NA	560 (8.5%)
Substance abuse services per year	NA	262 (3.9%)
Total children identified by any measure of superutilization	**12,332 (56%)**	**3726 (55%)**

Note: CBC is community-based care purchased services in Florida

Table 7.2 Types of superutilization and percentage among those identified as experiencing superutilization in the Tennessee sample

Class name	Description	Percentage of superutilization sample
Class 1	Foster care placement instability	23.0%
Class 2	Multiple foster care episodes	12.2%
Class 3	Child welfare service use	21.5%
Class 4	Duration in foster care	7.3%
Class 5	Medicaid outpatient service use	9.1%
Class 6	Medicaid emergency service use	8.6%
Class 7	Use of group/congregate care placements and high placement costs	18.1%
Total percentage of superutilization sample		**100.0%**

There were a number of similarities in findings across both study sites, though a few differences were also observed, reflecting variability in services data and service use across the sites. Both state samples had groups demonstrating superutilization regarding foster care placement instability, meaning those children with high number of placement moves, groups with children that had high

Table 7.3 Types of superutilization and percentage among those identified as experiencing superutilization in the Florida sample

Class name	Description	Percentage of superutilization sample
Class 1	Child welfare CBC purchased service use	14.0%
Class 2	Complex child welfare and Medicaid service use	5.4%
Class 3	Medicaid and mental health service use	23.2%
Class 4	Foster care placement instability	10.2%
Class 5	Multiple foster care episodes	19.9%
Class 6	Duration in foster care	5.6%
Class 7	Use of group home/residential treatment placements	8.8%
Class 8	Medicaid emergency services use	12.8%
Total percentage of superutilization sample		**100.0%**

Note: CBC refers to community-based care purchased services in Florida

numbers of episodes in foster care, groups with long durations in foster care, groups with high proportions of time in foster care in group homes or residential placements, and groups with high levels of other child welfare services. Both state samples also had superutilization groups that had high usage of Medicaid emergency services, but they showed some variability regarding groups with other Medicaid service use and, for the Florida sample, use of other mental health services.

While only the high-level findings are presented here, the full report (Weigensberg et al., 2018) has detailed descriptions of each superutilization group, including the characteristics of children in each group to better understand who the children are in each type of superutilization. For example, when looking at the group with high levels of foster care placement instability for the Tennessee sample, 55% of the children in this group have seven or more placement moves. Also, 92% of the children in this group received child welfare services, 71% received clothing assistance, 29% received substance abuse testing and treatment, and 23% received family or parenting support services. The largest percentages in regard to their reason for removal and placement in foster care were neglect (43%) and parental drug use (39%). Among the 62% that exited custody, almost half (48%) were reunified while about a quarter (24%) were adopted. This information about the characteristics of each of the superutilization groups was extremely helpful for the state agency partners to understand what types of children experienced the various types of superutilization.

7.2 Predictive Analysis Results

After identifying the various types of superutilization, we met with our state agency partners to brief them on the findings and identify which type of superutilization would be most useful to explore for predictive analysis. Both study sites wanted to understand predictive factors for children in the placement instability group to learn more about what factors at the time of entry are predictive of kids that may experience high numbers of placement moves.

In order to do the predictive analysis, we refined the study sample based on the data we had available. For each child in the sample, we established a 12-month period of time for our predictive analysis, starting at their date of entry into out-of-home care through up to 12 months in care. This predictive time period may have been shorter than 12 months for those who exited custody before then. We also used a 12-month look-back period, which included data available from 12 months prior to entry through the date of entry into out-of-home care. We also included a few additional data elements, even if they occurred before the 12 month period prior to entry if that was information that would typically be available to the caseworker at the time of entry, such as prior child welfare involvement with investigations or foster care episodes. With this refined sample and timeline, the predictive analysis explored what factors known at the time of entry into out-of-home placement are the most important predictors of placement instability superutilization within 12 months of entry.

We also measured placement instability superutilization using this 12-month predictive period of time. Using the 90th percentile threshold for superutilization of placement instability, as measured by number of placement moves, superutilization was operationalized as those children with five placement moves or more within 12 months of entry for the Tennessee sample and four placement moves or more within 12 months of entry for the Florida sample. We used 70% of the study sample for the training sample, and we used 30% of the sample as the testing sample for our predictive analysis.

Given the exploratory nature of our predictive analysis, we were as inclusive as possible with the data we had available for this study. The predictive models included the following types of variables where available for each study site: child demographics, prior child welfare investigations, reasons for removal and placement into foster care, types of foster care placement, child welfare custody episodes, child welfare services, Medicaid services, child welfare assessments, substance use and mental health assessments, and regional characteristics. We used a total of 65 predictor variables in the model for the Tennessee sample and 53 variables in the model for the Florida sample.

We examined three different types of predictive modeling approaches. We tested logistic regression with elastic net regularization, K-nearest neighbor, and random forest models and compared their overall predictive performance based on how well they performed on the training data (i.e., 70% of the full sample). We considered several model performance metrics, including model sensitivity, specificity, accuracy, positive predictive value, negative predictive value, and the area under the receiver operator characteristic curve (AUC), which was the primary metric we used for comparison. Higher values of this (typically above 0.70) are indicative of good predictive accuracy. The random forest model performed best on this criterion, with the AUC of 0.727 for the model with the Tennessee sample and 0.722 for the model with the Florida sample. Full results regarding our comparison of predictive models can be found in the final report (Weigensberg et al., 2018).

The results of the random forest model indicate the relative importance of the variables by showing how each contributes to the overall model fit. For random forest models, this can be determined by ranking individual predictors based on the mean percentage change in the Gini impurity index (James et al., 2013). This index measures the change in overall model fit that a given predictor contributes, with higher values indicating a greater contribution.

Figure 7.2 lists the eight most important predictors for the Tennessee sample based on the mean change in the Gini index. Based on the rankings, a child's age at entry into the first out-of-home placement during the prediction period is the most important variable by a wide margin (the mean decrease in the Gini index is over 350), followed by the number of prior investigations. Note that the next eight variables all appear to cluster together in terms of the change in the mean Gini index.

In the full study report, to aid in interpretation of findings, we provide the marginal predicted probabilities (partial density plots) of superutilization for the eight most important predictors based on their relative rank according to the Gini index. To summarize the key findings, for the Tennessee sample predictive model, the most important predictor is a child's age at entry into the first out-of-home placement during the prediction period. The predicted probability of experiencing superutilization based on placement instability is relatively steady for children at

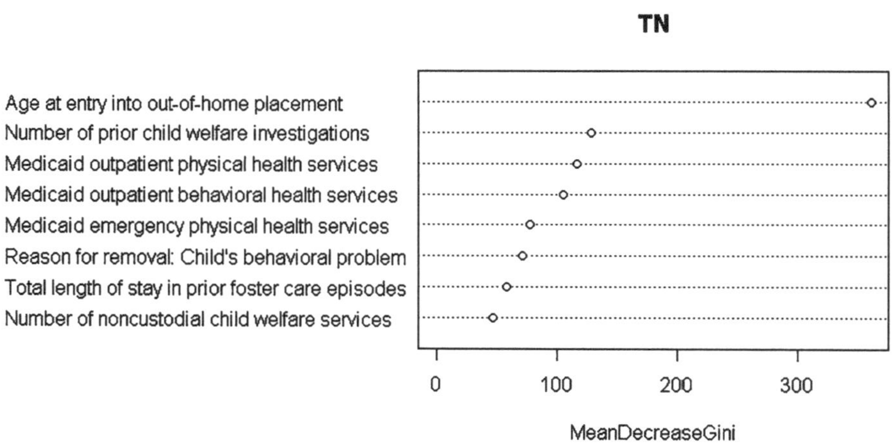

Fig. 7.2 Eight most important predictors for placement instability superutilization in Tennessee. (Source: Tennessee DCS; TennCare; American Community Survey 2015; Census 2010)

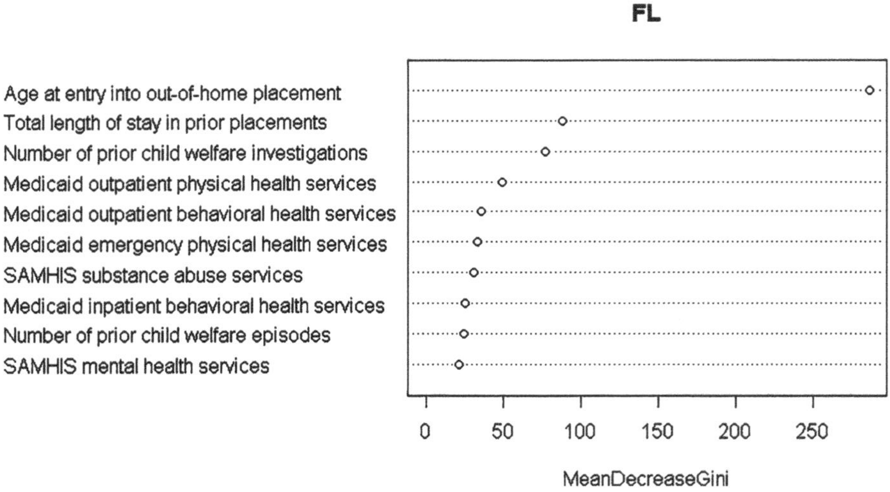

Fig. 7.3 Ten most important predictors of placement instability superutilization in Florida. (Source: Florida OCW; Florida AHCA data; Florida SAMH; American Community Survey 2015; Census 2010)

younger ages before increasing notably between ages 11 and 12, and children ages 15 and 16 have the highest probability of experiencing superutilization. The second most important predictor for the Tennessee model is the number of prior child welfare investigations. There appears to be a mild linear increase in the probability of experiencing high numbers of placement moves as the number of prior investigations increase. The next most important predictors all relate to prior receipt of Medicaid services for outpatient physical and behavioral health and emergency physical health. Although the magnitude of change in the Medicaid measures is not particularly high, the fact that these variables seem to have predictive strength suggests that it may be useful for child welfare agencies to know about Medicaid service history at the time of entry into custody.

Next, we looked at the results of the predictive model for the Florida sample using variable importance, as depicted in Fig. 7.3, and partial dependence plots, which are presented in the final report (Weigensberg et al., 2018). Figure 7.3 lists the 10 most important variables in the Florida model in terms of mean decrease in the Gini index.

Similar to the Tennessee findings, the most important predictor for placement instability in Florida is a child's age at entry into the first out-of-home placement during the 12-month prediction time period. The predicted probability of experiencing high numbers of placement moves in Florida increases notably between ages 11 and 12 and continues to increase with age. The second most important variable is the length of time spent in prior out-of-home foster care and the third most important variable is the number of prior child welfare investigations, which reinforce the importance of child's history in the child welfare system when identifying children at higher risk of superutilization.

Several of the most important variables involve Medicaid service receipt during the 12 months prior to entry into out-of-home care. In particular, as the number of Medicaid outpatient physical health services, outpatient behavioral health services, emergency physical health services, and inpatient behavioral health services increase, the predicted probability for high levels of placement instability increases. Taken together, these results suggest that it may be useful to share and incorporate Medicaid service data into early warning systems for child welfare, and comprehensive health histories should be included in any assessments or records at the time of entry into foster care.

We also found that use of state-funded substance abuse and mental health services during the 12 months prior to entry is predictive of placement stability superutilization in the Florida sample. If a child receives one or more non-Medicaid substance abuse or mental health service, his or her predicted probability of high levels of placement instability increases.

These findings suggest that it may be worthwhile to share and integrate data on substance abuse and mental health services with child welfare data to identify those most at risk of placement instability at time of entry into foster care.

It is also important to note that these variables may be impacted by other variables in the model in ways that are not yet well understood; that is, they potentially interact with others in the random forest model in a way that increases the predicted probability but is not easily detected by plotting the marginal probability on only the variable itself. Examining these dynamics may be beneficial for future work.

7.3 Conclusion

The research team engaged in this exploratory research with our site partners to better understand high levels of services use, including identifying types of superutilization and examining whether it was possible to predict superutilization of services at the time of entry into foster care. We assessed the benefit of bringing a variety of data sources together, including Medicaid, child welfare, and supplemental substance abuse and mental health services data for this purpose, and we explored how these linked data related to services could help inform what our site partners were doing to serve children and families.

Casey Family Programs has been actively engaged with both Tennessee and Florida to consider how to use the results from the study. In Tennessee, the information led the state to consider strategies to improve recruitment and retention of therapeutic foster care providers in order to reduce placement instability and the use of group care. The state also started to consider how they can share more information with their provider community. In Florida, the state expressed interest in further examining those children who had high levels of emergency services use, especially since many of these individuals were very young children. They also considered doing an intensive case review to help better understand these children and their medical needs. Finally, the state also expressed interest in fostering more collaboration between the state child welfare and Medicaid agencies.

Readers interested in learning more are encouraged to review the full report for the study (Weigensberg et al., 2018). The report includes rich technical information, as we were committed to providing thorough details and being transparent about our methods for others interested in pursuing similar research. We wanted to be sure others could understand our predictive analysis methods since many people view predictive analytics as a black box, given others are not often transparent about what data or methods are used in their modeling.

It is important to note this was an exploratory study. We learned many valuable insights regarding data linking across key service systems as well as understanding and predicting types of superutilization of services. We hope others can continue building on this research and advance the field to provide more needed insights regarding services for children in child welfare.

Furthermore, there is a growing need in the field to help respond to the opioid epidemic, given the increasing impact of substance use on children and families in the child welfare system. There is much potential to analyze child welfare data with Medicaid data and other behavioral health data to better understand how these systems engage families and provide treatment services. There could be much benefit to engaging families holistically from both the child welfare and behavioral health systems when trying to help families with substance use issues.

There is a tremendous opportunity to learn more by linking Medicaid and child welfare data. These data can provide important insights to learn about how to serve the needs of both children and parents. Linking these data together creates a broader view to look at services and outcomes across systems and identify ways to improve information sharing and service provision. With every service system facing limited resources and increased accountability, there is great potential to use linked cross-system data to identify more effective ways to serve families, not only to achieve positive outcomes for children and parents but also to provide services more efficiently in a coordinated manner.

Acknowledgments Dr. Weigensberg presented this study at the Child Maltreatment Solutions Network Conference on behalf of her study coauthors—Erin Maher, Peter J. Pecora, and Kirk O'Brien from Casey Family Programs as well as Derekh Cornwell, Lindsey Leininger, Matt Stagner, Sarah LeBarron, Jonathan Gellar, Sophie MacIntyre, and Richard Chapman from Mathematica. This study was conducted in collaboration with Casey Family Programs, whose mission is to provide, improve, and, ultimately, prevent the need for foster care. The presenting author thanks Linda Jewell Morgan and Fred Simmens of the Casey Family Programs' Strategic Consulting Team for their collaboration and work with the sites. She also thanks senior project advisors Toni Rozanski and Nadia Sexton. The project was made possible through the partnership and contributions of many people and organizations in the study sites. The presenting author thanks the study site partners, including the Tennessee Department of Children's Services, TennCare, the Florida Department of Children and Families, the Florida Agency for Healthcare Administration, and Eckerd Kids, the lead Community Based Care agency in Hillsborough, Pinellas, and Pasco Counties, for their engagement in the study. Lastly, this research would not have been possible without the contributions of many other Mathematica staff, including Christina Alva, Stephanie Barna, Daniel Kassler, Brenda Natzke, Jessica Nysenbaum, Matt Mleczko, Kelley Monzella, Nora Paxton, Liz Potamites, Dmitriy Poznyak, Christine Ross, Michael Sinclair, and Fei Xing.

References

James, G., Witten, D., Hastie, T., & Tibshriani, R. (2013). *An introduction to statistical learning*. Springer New York Inc.

Maher, E., Weigensberg, E., Stagner, M., Nysenbaum, J., & LeBarron, S. (2016). *Addressing unaddressed needs: Helping agencies target services to children and caregivers in child welfare. Issue brief.* Mathematica. https://www.mathematica.org/our-publications-and-findings/publications/addressing-unaddressed-needs-helping-agencies-target-services-to-children-and-caregivers-in-child

Weigensberg, E., Cornwell, D., Leininger, L., Stagner, M., LeBarron, S., Gellar, J., MacIntyre, S., Chapman, R., Maher, E., Pecora, P., & O'Brien, K. (2018). *Superutilization of child welfare, Medicaid, and other services. Final Report.* Mathematica. https://www.mathematica.org/our-publications-and-findings/publications/superutilization-of-child-welfare-medicaid-and-other-services

Chapter 8
Improving Child Welfare Practice Through Predictive Risk Modeling: Lessons from the Field

Rhema Vaithianathan, Stephanie Cuccaro-Alamin, and Emily Putnam-Hornstein

This volume is filled with descriptions of innovative methods for leveraging administrative data to help detect, prevent, and respond to child maltreatment. In this chapter, we focus on one such method: the use of predictive risk modeling (PRM) to help child welfare agencies more effectively identify children at risk of child maltreatment and to better target system responses to these children and their families.

PRM offers numerous distinct advantages over conventional risk assessment tools. One important advantage is the ability to identify latent risk, which can be difficult to detect during the acute crisis assessments that characterize front-end child welfare practice. In this context, this chapter attempts to illustrate how PRM has the potential to improve practice by supporting decision making at crucial phases such as at call screening. We also argue that by identifying children at risk of child maltreatment at birth, PRM can be used beyond simply supporting policy and practice decisions to enable proactive preventative engagement. Using case studies drawn from Douglas County, CO; Allegheny County, PA; and the country of New Zealand, we provide information regarding implementation, methodology, results, and any practice or policy changes that emerged from each use case. We close with important observations on the ethical uses of these powerful new methods.

R. Vaithianathan (✉)
Centre for Social Data Analytics, Auckland University of Technology, Auckland, New Zealand
e-mail: rhema.vaithianathan@aut.ac.nz

S. Cuccaro-Alamin
University of California at Berkeley, Berkeley, CA, USA
e-mail: stephca@berkeley.edu

E. Putnam-Hornstein
University of North Carolina at Chapel Hill, Chapel Hill, NC, USA
e-mail: eputnamhornstein@unc.edu

C. M. Connell, D. M. Crowley (eds.), *Strengthening Child Safety and Well-Being Through Integrated Data Solutions*, Child Maltreatment Solutions Network, https://doi.org/10.1007/978-3-031-36608-6_8

8.1 Latent Risk

Research consistently shows that humans are bad at making predictions regarding future events (Dawes et al., 1989; Kahneman & Tversky, 1973). For example, in a 2011 study regarding clinical prediction in health care, when trained clinicians in a hospital setting were asked "*which patients are more likely to be readmitted?,*" the performance of physicians, interns, case managers, and nurses at identifying who was at highest risk was no better a predictor than tossing a coin (Allaudeen et al., 2011).

Doctors and other health care clinicians often fail to correctly predict future events because as crisis service providers, they see patients at the peak of acuity, i.e., while they are at their worst. Such health care providers are, by definition, managers of acute episodes. Assessment during such episodes, however, rarely yields information regarding latent risk – the often invisible, but long-standing underlying factors that created the conditions for the emergence of the acute episode.

For instance, a clinician may assess the hospital readmission risk of a patient suffering a heart attack by considering only the prognosis for the cardiac event. However, for a patient with multiple latent risk factors such as other long-term health conditions or poor primary care, it may not be their heart disease but other health conditions that cause the next hospital visit. On the other hand, a patient presenting with a similar acute heart attack – but with no other long-term health conditions and therefore low latent risk – will have a much lower readmission risk. The problem in predicting future risk is that acute care clinicians only see what's immediately in front of them, and because latent risk is difficult to detect with acute clinical judgment alone, they are likely to predict the same long-term outcome (i.e., rates of readmission) for these very different patients.

This phenomenon is not restricted to health care; frontline child welfare workers face similar circumstances every day. As with health care clinicians, they move from one acute crisis to the next, making latent risk difficult to detect. For some families, all their risk factors might well be related to that which caused the acute episode that brought them to the attention of the child welfare services. On the other hand, families with complex and multi-faceted sources of risk and resilience can be harder to assess. Further complicating the workers' task is the fact that available risk assessment tools cannot reliably detect these latent factors. Instead they function like smoke alarms pointing workers to the fire. While child welfare workers are stretched, "running from fire to fire", we need to provide them with tools that more accurately identify latent risk – to slow them down, to take their time to effectively engage with families when they are not in crisis, and to hopefully address the long-term structural issues that are the apotheosis of preventive social work. Thus, in contrast, for health care workers, the majority of decision making is made with triage mentalities similar to emergency departments as opposed to primary care contexts. This chapter highlights the use of PRM methods for these purposes.

8.2 What Is Predictive Risk Modeling?

In child welfare practice, frontline workers utilize risk assessment tools to help guide case decision making. Structured Decision Making (SDM) and similar actuarial tools help provide decision support for case screenings. Typically, a child welfare hotline worker populates a paper form or online tool with information as it is gathered during a call. In addition to client demographics and other case characteristics, fields include information regarding prior maltreatment and known risks factors.

These tools then generate a score with weights/points given to specific risk factors. Weights are typically based on prior research regarding the relationship between the risk factors and adverse future events such as re-referral or out-of-home placement. The resulting summary score provides a metric that helps frontline workers assess risk for a specific child. The tools are often validated on large populations to ensure their validity and reliability.

Recent technological innovations associated with the availability of big data have increased interest in the application of PRM as an alternate risk assessment strategy, to SDM and actuarial tools, in child welfare services. PRM is an automatic risk scoring method which can be applied to routinely collected administrative data, to calculate a risk score for a wide range of adverse events. From a practitioner standpoint, PRM can be thought of as a risk assessment tool where all the fields are "pre-filled" using historic data. Importantly, PRM tools are constructed and validated on the population on which they are being used – therefore, the PRM fields used across jurisdictions may be very different from one another.

When applied responsibly, PRM is able to overcome many of the limitations of conventional risk assessment tools. For example, the number of fields available for risk scoring is not limited by the child welfare worker's willingness and ability to complete the fields. Instead, the PRM is able to include hundreds of fields garnered from administrative data sources such as the existing Statewide Automated Child Welfare Information System (SACWIS). Additionally, unlike actuarial tools that are often validated only once on a single population at a specific time, PRM utilizes data specific to the population of interest so that weights are customized and validated for that specific group. Unlike out-of-date paper tools, the quality and the predictability of PRM can also be monitored in real time.

In addition to these benefits, PRM offers something that paper tools are unable to provide. Specifically, it serves as the "smoke alarm" for the latent risk that often goes unidentified during crisis assessments. Furthermore, PRM is able to provide workers with the indication that in the long-term – even if the current crisis abates – the child or family is at risk of chronic involvement in Child Protective Services (CPS) and, therefore, preventive services should be targeted to them now. Ultimately, the identification of latent risk with PRM moves child welfare practice closer toward a goal of primary prevention.

PRM is by no means perfect. Because it relies on routinely collected administrative data, information quality can impact reliability and validity of risk estimates. For instance, important risk and protective factors might not be represented and the data can reflect systemic biases present in agency practice (e.g., racial and/or ethnic disproportionality). Over time, however, PRM can help identify these biases, enabling agencies to improve practice and information quality.

Social work best practice takes place when workers can use professional judgment to assess important protective factors such as parental interaction, or mother and child warmth, as well as risk factors. Unfortunately, instead of providing frontline workers with the time to conduct these critical assessments, the current systems burden them with mandatory forms and requires them to do high-risk mathematical calculations. PRM leverages technology to access available data to calculate summary risk scores, providing workers with more opportunity to focus on practice and supporting families.

8.3 Case Studies

This section reviews case studies from Douglas County, CO., and Allegheny County, PA., in the United States and case studies from New Zealand to illustrate how PRM has been used in this context to improve practice, to support decision making, and to identify children at risk at birth.

8.3.1 Improving Practice

PRM can be applied to improve the critical decision of who to screen-in following a referral for abuse and neglect. In the United States, around 1 in 3 children are investigated for abuse and neglect by the time that they turn 18 years (Kim et al., 2017). Indeed, every time there is a death or high-profile abuse case, the number of referrals increase, both through public awareness and explicit legislative changes. For example, following the 2011 child sexual abuse scandal at Penn State University, Pennsylvania lawmakers updated the state's Child Protective Services Law. Changes included lowering thresholds for qualifying injuries classified as child abuse and neglect, expanding the list of mandated reporters and training requirements for these individuals, and adding an online reporting system as well as other important changes designed to improve child safety. These sorts of reforms will often increase the number of hotline calls, which in turn increases the number of potential families who are subject to investigations. The system is then flooded with calls, making it crucial that screening decisions are correct.

This scenario has been demonstrated in Douglas County, CO, which is part of the Denver-Aurora-Lakewood, CO Metropolitan Area which has seen rapid population growth. In 2016, the county child welfare agency sought the assistance of academic researchers to pilot a PRM method to help better understand their front-end screening decisions. From the outset, the agency prioritized transparency and has shared its experiences with stakeholders and the larger research community.

Using linked child welfare and public welfare data, researchers built an algorithmic model to predict out-of-home placement within two years of referral. For every child-level referral received, the model weighted two years of prior history, including hundreds of factors related to the child under referral, and any sibling under referral, as well as parents. With every prior referral in the two-year window scored, children were divided into risk ventiles (1–20).[1] A score of 1 indicated a low probability that the child would end up being placed within two years, whereas a score of 20 meant there was a very high probability that the child would be placed out-of-home within two years' time.

Researchers tested the accuracy of the tool by following the high- (score = 20) and low- (score = 1) risk groups and examining their actual child welfare outcomes over the 24 months following their CPS referral. The test data used for these analyses had not been used to build the initial PRM. Results showed that, for this population almost half (45%) of the high-risk children ended in a placement within two years. By comparison, less than one and a half percent of children coded as low risk were placed within two years.

Although the algorithm was accurate, screening decisions are not about preventing within-two-year placements; rather, they focus on identifying children who are in need of services or protection. In the PRM context, placement serves as a proxy for latent risk – it reflects the likelihood that in the next two years, child welfare agency staff will be so concerned about a given child that they will ask a judge to authorize the child's removal from their current living situation.

Researchers therefore tested whether child placement was a valid proxy for long-term risk by analyzing the risk profiles of fatalities and near fatalities in the state during the 20 days prior to the critical incident. Colorado had a total of 120 fatality or near-fatality cases for children under 10 years old that could be identified using the Colorado SACWIS (referred to as TRAILS) for the period under investigation. Of these, 42 had a child abuse and neglect referral prior to the critical incident. Researchers calculated the maximum score the algorithm would have given all referrals received 20 days prior to the critical incident and results showed that 25 out of the 42 fatality and near-fatality victims would have been scored as high risk (18–20 range) and only 1 of the 42 would have been scored as low risk (1–4 range). Using fatality and near-fatality data helped verify that the placement outcome was in fact a valid proxy for latent risk.

Based on these findings, researchers then examined to better understand the agency's practices with respect to screening decisions and subsequent child welfare contact for children by risk quintile. In total, 38% of children identified as high risk by the algorithm (score = 20) were screened out at the time of referral. Twenty-seven percent of these screened out children were later re-referred and placed within one year, and 35% were re-referred and placed within two years. At the other end of the spectrum, 65% of children scored as low risk by the algorithm (score = 1) were

[1] What we mean by ventiles is that the scores are calibrated so that roughly 5% of children receive each score from 1 to 20. In other words, the least risky 5% would be allocated a score of 1, and the most risky 5% are allocated a score of 20.

screened in and one percent were placed within two years. The findings, which have been replicated in both California and Allegheny County, Pennsylvania, suggest that screening rates are relatively flat and often do not reflect latent risk levels. Workers screened out a significant portion of referrals scored as high-risk and screened in a significant portion of low-risk referrals.

Because of the safety risks to children, Douglas County, Colorado, conducted detailed case reviews to better understand cases with type II errors (i.e., where high-risk cases were screened out). The review suggested that a common factor in these cases was decision fatigue. For example, in situations where multiple calls were received for a child, if investigated and determined to be as low risk, subsequent calls were often ignored. The high frequency of calls caused the tool to score the case as high risk, but caseworkers who had already been out once were understandably reluctant to reassess the case. The analysis using PRM helped caseworkers make important discoveries about their practice. Before implementing the PRM live at the front-end, Douglas County, CO, decided to conduct a randomized control experiment. This analysis is described in Chap. 4 (Putnam-Hornstein, Cuccaro-Alamin, & Vaithianathan, This Volume).

This experience with PRM by Douglas County shows that even before the tool was deployed, it helped illuminate weaknesses in current practice. As the model's primary function is to "summarize and weigh" the historical information in a consistent way, critics of PRM sometimes dismiss the approach because resulting models cannot predict exactly what will happen for a specific child. It is important to recognize that, although such models are far from perfect, when it comes to child safety, the field cannot let perfect be the enemy of good. As the case study of Douglas County illustrates, PRM *can* improve practice by helping direct attention to cases with high latent risk. With the current front-end assessment paradigm, we are effectively playing dodgeball in the dark – we don't know from what direction the ball will come. PRM is like turning the lights on – you can see the pitch, and although the ball might not come from the same direction every time, it tells you which direction to face.

8.3.2 Supporting Decision Making

Jurisdictions such as Allegheny County, PA, are using PRM to investigate ways to improve their front-end screening decisions. Allegheny County is unique in that it has spent decades building an integrated administrative data system for their public service programs. Launched in 1999, the Department of Human Services (DHS) Data Warehouse is integrated at the client level and includes data from programs which include child welfare, drug and alcohol, criminal justice, mental health, education, housing/homelessness, and income support. The Data Warehouse has resulted in strong community acceptance of a 360-degree view of clients and greater transparency around data-driven policies.

Using this integrated data system, the county built the Allegheny Family Screening Tool (AFST), a child welfare decision support algorithm for front-end child welfare processes (Allegheny County Department of Human Services, n.d.). Every time a call comes into the child maltreatment hotline, the model analyzes hundreds of historical data elements to predict the likelihood of the child on a referral experiencing a foster care placement within 24 months. The AFST calculates the risk score, of between 1 and 20, for this outcome. Scores between 18 and 20, as well as cases that meet other criteria including a maximum age, and home-school status, are flagged as mandatory screen-ins. Despite the terminology, these mandatory screen-ins can be (and frequently are) over-ridden by the call screening supervisors. For other cases, the score is utilized as part of the risk assessment process.

In designing and training the AFST, placement serves as a proxy for more objective levels of harm. Because Allegheny County is too small to validate risk scores using fatality and near-fatality data, researchers instead used subsequent hospitalizations, as Allegheny County has integrated health care records into their linked data systems. Specifically, the researchers matched scored referrals to hospitalization records from Children's Hospital in Pittsburgh. Results showed that high-risk children were 17 times more likely than low-risk children to be hospitalized for physical assault and 21 times more likely to be admitted for self-inflicted injury or suicide attempt during adolescence (compared to low-scored children).

Before a formal evaluation of the AFST implementation was completed, researchers examined historical decision making by examining subsequent out-of-home placements by risk quintile. Prior to the AFST implementation, nearly half (48%) of the county's lowest risk cases were screened in, but only 1.4% of these went on to experience an out-of-home placement. Among the children who scored 20, around one-third were screened out, and yet 30% of these screened out children were later re-referred and placed within one year. An independent impact evaluation of the AFST implementation was completed by researchers at Stanford University in March 2019 (Goldhaber-Fiebert & Prince, 2019). Results showed that use of the tool increased the accurate identification of these high-risk children who needed further intervention services and that this practice improvement came without adverse consequences and without increasing the workload on investigators. Crucially, the tool also reduced racial disparities in case-opening rates by slightly reducing screening in of Black children and increasing the case-opening rates of high-risk non-Black children.

PRM risk stratification has also highlighted differences in screening decisions by race/ethnicity. Results from initial risk stratification during the AFST development showed that a large proportion of Black children who were being screened in were low-risk children. Conversely, a lot of high-risk White children were being screened out. PRM helped illuminate that the county might be over-surveilling Black children and under-protecting White children. The formal AFST evaluation found that the tool implementation resulted in reductions in racial disparities in case openings following investigations. Specifically, it found modest but statistically significant increases in cases opened for White children coupled with declines in the rate at which Black children were screened in for investigation (Goldhaber-Fiebert & Prince, 2019).

While proven useful, critics of this type of real-time PRM implementation raise concerns that rather than supporting decision making, this approach may end up driving the decisions and eventually crowding out good social work practice. The concern is a valid concern. To counteract this during implementation, Allegheny County required that workers with access to PRM risk scores receive training to understand how the algorithm works and other aspects of practice with the risk model such as what to do if the referral is flagged as a mandatory screen-in. Importantly, this training involves educating workers on the model's false positives – cases where the model scored a child as high risk, but after two years of follow-up, there was no re-referral or placement. While early on, the research team experimented with using techniques such as Shapely values to provide insight into why certain children score high or low, staff generally did not find this useful. Therefore, training was kept high-level aimed and keeping workers engaged with using the PRM.

In addition to consistent worker training, the county has been diligent with regard to continuous quality improvement. Specifically, on a monthly basis, a report is sent to the leadership team summarizing screening decisions stratified by risk level. Results suggest that there are still many high-risk cases being screened out and low-risk cases being screened in. Allegheny County stakeholders see this as evidence that the tool is operating effectively by alerting workers to potential risk, but not having it override their immense training and expertise to effectively work each case.

8.3.2.1 Ethical Considerations

While PRM has resulted in improved accuracy for screening decisions in Allegheny County, the experience highlights the importance of having an ethical framework to guide adoption and implementation. Early on, Allegheny County engaged a team of researchers to conduct an ethical analysis of PRM use cases (Dare & Gambrill, 2017). The review covered issues related to consent, disclosure of confidential information, false negatives and false positives, stigmatization, and racial disparity, as well as issues related to professional competence and training, provision and identification of effective interventions, ongoing monitoring, and resource allocation. The authors made several important recommendations and concluded that "subject to the recommendations in this report, the implementation of the AFST is ethically appropriate. Indeed, we believe that there are significant ethical issues in not using the most accurate risk prediction measure (p.9)".

In addition to adopting an ethical framework for guiding implementation and continuous worker training, the Allegheny project has shown that high levels of transparency and community engagement are required for successful implementation. In particular, it is imperative that information about how the tools work be shared transparently with those most likely to have it applied to them. For instance, imagine a client with history in the administrative data systems who the algorithm might score as high risk, but who has done the important work of recovery and is trying to reunite – they may worry the social worker will still intervene because of high-risk scores. This underscores why mandatory or imposed interventions should never be considered in response to PRM. Algorithms must support rather than supplant practice and clinical decision making. Building community trust around PRM is difficult work, but Allegheny County has been a leader in this regard.

8.3.3 Identify At-Risk Children at Birth

Using New Zealand data, a prototype tool was built to prospectively identify children at risk for research purposes. The country has an incredibly rich integrated database, known as the Integrated Data Infrastructure (IDI), available for research. The IDI includes de-identified linked micro-level data about people and households (Stats, 2018). Using the IDI, researchers set out to determine whether they could build prospective PRMs to predict subsequent maltreatment for an entire birth cohort. Using linked health, welfare, child welfare, and criminal justice data, researchers built a model to predict a child's risk of a substantiated maltreatment by age 5. When validated, results showed that half of the children in the high-risk decile experienced substantiated maltreatment by age five years compared to only 2% of the children in the lowest risk decile (Ministry of Social Development, 2014).

Using similar data, researchers later expanded the study to see whether these high-risk children were also at elevated risk for injury and mortality in early childhood. For all children born in New Zealand in 2010 and 2011, researchers built a PRM to predict a child's risk of a substantiated maltreatment by age two (Vaithianathan et al., 2018). Children in New Zealand's home-visiting program called "Family Start" were excluded. Children were then risk stratified into deciles and flagged as "very high risk" if they were in the top 10% of the score distribution for maltreatment and "high risk" if they were in the top 20%.

The study then compared injury and mortality rates for "very high-risk" and "high-risk" children and the remainder of the birth cohort by age three. Stratified outcomes included inflicted injury deaths, unintentional injury deaths, and sudden unexpected infant death (SUID) as well as hospitalization. Overall, children scored as high risk and very high risk for substantiation were also at greater risk for other adverse outcomes. Compared to other children, they had much higher post-neonatal mortality rates (4.8 times and 4.2 times greater, respectively). Most importantly, with regard to the specific outcomes examined, children classified as high risk were almost 10 times more likely to experience unintentional injury (9.9 times), post-neonatal inflicted injury death (9 times), sudden unexpected infant death (SUID) (8.5 times), and hospitalization for a maltreatment injury (9.6 times) each by age two years.

8.3.4 Program Redesign

PRM can be employed to effectively redesign and refocus prevention programs. Research has shown that home-visiting programs are effective interventions in addressing child maltreatment and mortality risk (Eckenrode et al., 2017; D. Olds et al., 1998; Olds et al., 1997). An evaluation of New Zealand's home-visiting program "Family Start" showed that it resulted in reduced infant mortality deaths

(Vaithianathan et al., 2016). Researchers used PRM to examine whether this important intervention was being effectively targeted to high-risk families.

To do so, researchers linked PRM scores with home-visiting records in the IDI and found that despite being funded to target children in the highest five percent of risk, the majority of the children receiving "Family Start" home-visiting services were not in the highest risk decile (Vaithianathan et al., 2018). As a result of this research and studies which replicated the analysis within the ministries, policy makers have been able to redirect New Zealand's home-visiting resources toward higher risk families.

As this chapter has illustrated, PRM analyses can help us understand a lot about child maltreatment risk, but they are not designed to tell us what factors need to change in order to prevent it. More in-depth research is required to help identify solutions. Having predicted the risk of maltreatment for children at birth using a PRM, New Zealand researchers are now exploring the qualitative factors that must be addressed to design preventive interventions (Walsh et al., 2019).

Using the "Growing Up in New Zealand Study," a longitudinal cohort study, researchers built a PRM at birth to predict the likelihood a child would experience Adverse Childhood Experiences (ACES) during the first five years of life. They prospectively followed children in the two highest risk deciles (scores of 9 and 10) to determine how many ACES they actually experience. The average was 1.8 ACES. Interestingly, a number of high-risk children did not experience any ACES. For this group of children who "beat the odds," researchers then examined over 750 survey factors relating to community (36%), family finance (23%), parent/child interactions (18%), parent health (40%), and mother/partner interactions (nine percent).

Using data mining techniques, researchers isolated factors protective for experiencing ACES. Mother/partner interactions followed by parent health and wellness were most protective, while community and neighborhood factors accounted for only three percent of protective factors. These findings are important for practice as home-visiting programs pay very little attention to partners, but the quality of the relationship may prove protective. This important finding could not emerge from PRM alone; linkage to rich survey data was required for this work.

8.4 Conclusion

As the case studies of Douglas County, Allegheny County, and New Zealand demonstrate, whether used in practice as a decision support tool, used in academic research to better understand risk factors, or used by program developers to aid in redesigning programs to ensure risk factors are effectively targeted, PRM provides distinct advantages for risk assessment.

PRM's ability to identify latent risk provides new opportunities to detect, prevent, and respond to child maltreatment. Application of PRM can improve practice by alerting frontline workers to risk that often goes undetected when they are acutely focused on immediate safety. These latent factors are critical for prevention.

Although PRM is not a perfect solution, it does provide much needed direction in the risk assessment process, not to supplant decision making but to support it. Again, where child safety is concerned, perfect should not be the enemy of good.

Concerns among stakeholders that the use of such tools might exacerbate bias or undermine clinical decision making are, and will continue to be, challenging to overcome. The approach and evidence outlined in this chapter might provide comfort to these groups that with thoughtful implementation and ongoing monitoring and evaluation, the use of PRM tools can be a valuable aid for child welfare systems. Regardless of the use case, elements of the responsible and successful implementation of PRM include agency leadership, independent ethical reviews, external validation, examination of model fairness/disparities, clarity and transparency of data use, internal (agency) and external (community) engagement, continuous communications, and media staff training, as well as continuous evaluation and monitoring.

References

Allaudeen, N., Schnipper, J. L., Orav, E. J., Wachter, R. M., & Vidyarthi, A. R. (2011). Inability of providers to predict unplanned readmissions. *Journal of General Internal Medicine, 26*(7), 771–776. https://doi.org/10.1007/s11606-011-1663-3

Allegheny County Department of Human Services. (n.d.). *Developing predictive risk models to support child maltreatment hotline screening decisions*. Retrieved from Allegheny County Analytics website: https://www.alleghenycountyanalytics.us/index.php/2019/05/01/developing-predictive-risk-models-support-child-maltreatment-hotline-screening-decisions/

Dare, T., & Gambrill, E. (2017). Ethical analysis: Predictive risk models at call screening for Allegheny County. *Allegheny County Analytics*.

Dawes, R. M., Faust, D., & Meehl, P. E. (1989). Clinical versus actuarial judgment. *Science, 243*(4899), 1668–1674. https://doi.org/10.1126/science.2648573

Eckenrode, J., Campa, M. I., Morris, P. A., Henderson, C. R., Bolger, K. E., Kitzman, H., & Olds, D. L. (2017). The prevention of child maltreatment through the nurse family partnership program: Mediating effects in a long-term follow-up study. *Child Maltreatment, 22*(2), 92–99. https://doi.org/10.1177/1077559516685185

Goldhaber-Fiebert, J. D., & Prince, L. (March 20, 2019). *Impact evaluation of a predictive risk modeling tool for Allegheny County's child welfare office*. Stanford University.

Kahneman, D., & Tversky, A. (1973). On the psychology of prediction. *Psychological Review, 80*(4), 237–251. https://doi.org/10.1037/h0034747

Kim, H., Wildeman, C., Jonson-Reid, M., & Drake, B. (2017). Lifetime prevalence of investigating child maltreatment among US children. *American Journal of Public Health, 107*(2), 274–280. https://doi.org/10.2105/AJPH.2016.303545

Ministry of Social Development. (2014). *The Feasibility of Using Predictive Risk Modelling to Identify New-Born Children Who Are High Priority for Preventive Services.*. Retrieved from https://www.msd.govt.nz/documents/about-msd-and-our-work/publications-resources/research/predictive-modelling/00-feasibility-study-report.pdf

Olds, D. L., Eckenrode, J., Henderson, C. R., Kitzman, H., Powers, J., Cole, R., et al. (1997). Long-term effects of home visitation on maternal life course and child abuse and neglect. Fifteen-year follow-up of a randomized trial. *JAMA, 278*(8), 637–643.

Olds, D., Charles R. Henderson, J., Cole, R., Eckenrode, J., Kitzman, H., Luckey, D., et al. (1998). Long-term effects of nurse home visitation on Children's criminal and antisocial behavior: 15-year follow-up of a randomized controlled trial. *JAMA, 280*(14), 1238–1244. https://doi.org/10.1001/jama.280.14.1238

Stats NZ. (2018, July 1). Integrated Data Infrastructure. Retrieved from https://www.stats.govt.nz/integrated-data/integrated-data-infrastructure/

Vaithianathan, R., Wilson, M., Maloney, T., & Baird, S. (2016). *The impact of the family start home visiting Programme on outcomes for mothers and children: A quasi-experimental study.* Retrieved from Ministry of Social Development website: https://www.msd.govt.nz/documents/about-msd-and-our-work/publications-resources/evaluation/family-start-outcomes-study/family-start-impact-study-report.docx

Vaithianathan, R., Rouland, B., & Putnam-Hornstein, E. (2018). Injury and mortality among children identified as at high risk of maltreatment. *Pediatrics, 141*(2), e20172882. https://doi.org/10.1542/peds.2017-2882

Walsh, M., Joyce, S., Maloney, T., & Vaithianathan, R. (2019). *Protective factors of children and families at highest risk of adverse childhood experiences: An analysis of children and families in the Growing up in New Zealand data who "beat the odds."* Retrieved from Ministry of Social Development website: https://www.msd.govt.nz/documents/about-msd-and-our-work/publications-resources/research/children-and-families-research-fund/children-and-families-research-fund-report-protective-factors-aces-april-2019-final.pdf

Chapter 9
Diverse Perspectives on the Promise and Challenge of Child Welfare Data Integration: Panel Discussions from the Practice and Research Communities

Christian M. Connell and Jennie G. Noll

The chapters in this volume reflect the broader proceedings of a 2-day conference sponsored by the Child Maltreatment Solutions Network at the Pennsylvania State University on "Strengthening Child Safety and Wellbeing Through Integrated Data Solutions," in the fall of 2018. The final session of the conference included two separate panel discussions reflecting a diversity of perspectives on the promise and challenge of using integrated administrative data sources to promote child safety and well-being through public systems. The first panel was led by Dr. Jennie Noll, former director of the Pennsylvania State University Child Maltreatment Solutions Network and current director of the university's Center for Safe and Healthy Children. The panel provided a forum for open discussion from child welfare administrators from state and county systems about their experiences using system data to carry out research and evaluation activities to improve system practices for children and families in their communities. Audience members were encouraged to share their experiences "from the field" or to pose questions for conference presenters and attendees. The second panel was led by Dr. Christian Connell, current director of the Child Maltreatment Solutions Network. This panel involved a more directed conversation with conference presenters, including chapter authors as well as other session presenters, that also provided an opportunity for audience reflection on the 2-day conference. The summaries below reflect an abstraction of the broader

C. M. Connell (✉)
Department of Human Development and Family Studies & Child Maltreatment Solutions Network, The Pennsylvania State University, University Park, PA, USA
e-mail: cmc128@psu.edu

J. G. Noll
Department of Human Development and Family Studies, Center for Safe & Healthy Children, College of Health and Human Development, PI, NICHD P50 Capstone Center of Excellence, MPI, NICHD T32 Training Grant, The Pennsylvania State University, University Park, PA, USA

C. M. Connell, D. M. Crowley (eds.), *Strengthening Child Safety and Well-Being Through Integrated Data Solutions*, Child Maltreatment Solutions Network, https://doi.org/10.1007/978-3-031-36608-6_9

panel discussions, rather than verbatim transcripts of these sessions. Instead, this summary reflects the tenor of the discussion with respect to both questions and responses—the summaries, therefore, reflect the interpretations of participants, rather than a word-for-word reflection of speakers' thoughts and ideas. Names or other identifiers of panel participants and audience members have been removed to maintain anonymity of the open conversation.

9.1 Part I: Perspectives from the Field: An Open Discussion with Child Welfare Administrators

>> **Moderator (MOD)** We want to have a focused discussion on learning more from people in the field to get a sense of community priorities or innovative opportunities for research and system partnerships and engagement. A significant aspect of our center is focused on community-engaged research. Since we started, I and other faculty involved have been talking to county Children and Youth administrators, caseworkers and social workers, providers, and leaders of larger agencies and asking questions. We want to know their perspectives on how research can be helpful to them, the types of research and policy questions they have, or whether research activities can answer important questions about the populations they work with.

This conference provides a unique opportunity with children and youth workers from a number of states to share their reflections on the challenges using administrative data. We've talked a lot about using administrative data algorithms for safety and risk assessments, but we want to know about other applications for use of administrative data. We also want to hear questions that you have about this type of work, or about issues that you would like to be able to use data to address. Finally, we want to hear your perspectives about some of the challenges around liability—issues of trust, security, or other barriers to working with these data.

>> **Audience 1** I serve as a Deputy Director for a large urban human services agency. Among other projects, we focus on child welfare and adult protection. About 4 years ago, our state auditor requested a statewide workload study that was a huge driver for some funding. The study also proposed an ideal caseload ratio, which led State officials to identify workforce needs and develop strategies to work toward hiring goals. This initiative has had a huge impact on the research and policy that we're seeing "on the ground," and it has provided valuable information to our local agencies as a result.

I also want to highlight a couple of issues that relate to the use of administrative data for predictive analytics and estimation of risk scores. As mentioned, some have expressed concern about the sensitivity of those kinds of scores, but I also see the use of these methods as an opportunity to engage families. There could be a role for a case worker to help families better understand the meaning of these types of risk scores and the implications for accessing necessary or useful services and supports. Policies could be developed to better support case workers or administrative staff in reviewing risk scores and score drivers with families, and families could meet with a "system navigator" to help them connect with resources to help reduce risks that may have contributed to elevated scores.

>> **Audience 2** I'm a director for a county-level Children and Youth Services office. I think one of the challenges to using child welfare data for research and analyses is the lack of standardization in some systems. Our office subcontracts a significant volume of the direct hours spent with families to external (i.e., community) providers. While some of the information about these contacts does get back into our official case file, it is rarely standardized and integrated into the system our caseworkers use. Most often, the information comes into our system in the form of a scanned report (i.e., a PDF or similar document) that is linked to the case record. Those reports include an incredible amount of information about what's happening through these contacts—the types of risk factors present in households, the services being provided, how things are improving—but it is not clear how we can get this information standardized and better integrated into our system in a format that makes it easier for our case workers and also makes it usable from a research and analysis perspective. We need solutions for this type of issue.

>> **Audience 3** Related to this concern, a lot of regional agencies are in the process of redesigning their data systems to enhance their electronic records. Agencies often tend to "circle the wagons" around their own agency and their own needs, but it may be useful for state partners (or multiple such agencies) to convene meetings that involved multiple agencies or providers around the table. If we were able to discuss some of these concerns while we were in the process of redesigning, or constructing, our data systems, we might be able to better plan for issues of inter-operability to facilitate information sharing across such systems.

>> **MOD** That's a good suggestion. Are community-based providers keeping the kind of data that child welfare agencies would find useful in terms of documenting information about the types of services provided, and is it accessible to child welfare agencies?

>> **Audience 4** I'm a senior research officer in the child welfare agency of a more urban county. I think all of our systems are in different places in terms of how we collect, utilize, and integrate data. Our county office has several layers of providers. In the traditional area of child protection or child welfare operations, our system is more integrated, but we also have divisions of juvenile justice and prevention that are not as integrated into our system, so that is a challenge.

It has been really encouraging to hear the presentations and to think about strategies to foster research partnerships. I think a lot of areas, especially larger agencies and jurisdictions, are investing in county-based research teams. There has been a lot of investment in our county, over time, to develop a division of performance management and to have a smaller data and research team within that organizational structure. Research partnerships can be useful in many different ways, but one important benefit is in building the infrastructure for data-driven decision-making within counties. These types of teams boost the internal capacity of counties and agencies to conduct research without relying on external capacity in terms of research skills or knowledge to carry out this type of work. That doesn't mean there is not a role for external partnerships, but I think it is critical to think about how partnerships can help to build that infrastructure—including conducting trainings with county teams about how to use administrative data for various research purposes. We also think about starting research partnerships earlier on in the process and about how partners can assist in the conceptualization and design of the study together with agency staff—as opposed to forming partnerships when the studies are in their final form and there isn't much of an opportunity to adjust these aspects. Finally, we are thinking about the use of integrated data and studies more broadly to produce actionable results and about what it means to implement research that actually makes a difference in our agency practices and policies.

>> **Audience 5** I am the director of a mid-sized county Children and Youth Services agency, and we don't have a senior researcher; we don't have a junior researcher, for that matter (laughter)! I think that is fairly typical, since we are certainly not the smallest county system in our state. I think there needs to be more statewide collaboration among private providers and among county agencies that are not as well staffed as larger agencies to be able to implement the kinds of research activities discussed at this conference. In addition, we need to identify strategies to include child welfare staff in these activities. Finally, we need to look for ways to better involve or engage members of our communities in these types of activities. I believe it is critical that agencies are more transparent with community members and other stakeholders in any activities that make use of the agency's—and I would argue it's actually the community's—data.

>> **Audience 6** I work for our state's child welfare agency, and we have been partnering with academic researchers as part of the work that they are doing to look at children's outcomes. I think part of what we want to do is continue to build on these types of relationships and look at ways to improve our interactions or find new and better ways to work together. I wanted to comment on the federal requirements being promulgated for Comprehensive Child Welfare Information Systems (CCWIS). Our state is one that has been approved to move forward under these new requirements, and some of the requirements will help us do a better job coordinating research activities that use the information systems that are developed. The requirements call for data standards and for data exchange standards. They also call for agencies that are contributing data to the overall child welfare system to comply with standards and common data exchanges. Many of the provider agencies in our state may be considered "child welfare contributing agencies"—the term that they're using. I think our challenge has always been that we have numerous counties and our state has multiple case management systems. There are counties that are

looking into doing more of this type of research, and they are looking for evidence-based ways to improve child welfare outcomes. I think we need to identify ways for counties to share what they're learning, because we have some larger counties who are able to do that type of work, but other counties who really aren't doing any type of research.

>> **MOD** Great, thank you for sharing. We have time for one final comment.

>> **Audience 7** Thank you. I am the director of a large county system in our state, and I think it would be very useful to organize a meeting with other administrators and county staff to identify the key variables we need in our system, including those from community providers, to make sure we are leveraging money and resources most appropriately. We need to know that when we spend millions of dollars in each county in services, we are spending that money in the most effective and efficient way possible. I'm challenged by that concern every day with my county commissioners. Our agency contributes a significant amount of money to our state match, and I want to know that I am investing in and using the most effective programs and services, and that we are getting the right data and research to inform these types of decisions. Having us come together to start that process and to use more data-driven decision-making is where we need to go in the future from a budgetary standpoint and to ensure that services make a difference.

>> **MOD** That is an important point that could be the focus of future meetings and discussions—how to leverage data-driven decision-making to inform practice and policy within the budgetary constraints of the child welfare system.

Next, we want to shift to our next panel involving some of the presenters who have shared their work during this conference over this past couple of days. This is a really unique feature of our Network conferences—bringing together applied researchers and experts, arguably among the leaders of this type of work across the country and internationally, to share their perspectives and engage in this type of discussion.

9.2 Part II: Perspectives of Applied Researchers: Key Themes and Future Directions

>> **MOD** Some of broad themes of the conference included access to systematic or high-quality data, integration of data from different sources, limited information on service access and delivery, and the need for better linking between indicators of risk and service needs. Collaboration was also a frequent theme—including boosting internal agency capacity, promoting academic and public agency collaborative efforts, and promoting cross-system collaboration. A focus of the conference was to set priorities for a national agenda to facilitate use of public administrative data

systems to promote child and family safety and well-being. What do presenters see as some of the key challenges to setting this type of agenda, and what are your visions for the field?

>> **Presenter 1** I think that there are two things that may facilitate a partnership between state or local agencies and the research community. One is to try to reduce the amount of transaction costs—the number of times that we have to make data requests or go through different processes—so that we are able to sustain this work for a longer period of time (i.e., where it's the same data that you're going to get on a regular basis, or where you have regular access to the data). Reducing those sorts of transaction costs will make it easier for everyone involved. You can make one set of security protocols, one set of data use agreements, and fill in the research questions as you go.

Second, we need to recognize the needs of both parties in our research collaborations. One of the things I admire about the academics that have spoken at this conference is how much they understand about the actual programs and systems that they are studying. Practitioners and those working in state or county agencies should expect that folks are not just going to use their data to go and study their own research questions and just talk to other academics. Practitioners should expect that researchers will use their data to help improve the agency's systems and the lives of the families they work with. That said, it is also important to understand that academic researchers are also incentivized to speak to the broader scholarly field. It is important to find the balance between understanding that academic researchers have the responsibility to use your data in a way that really helps improve your systems and advances the understanding and practice of your setting, but also to be okay with the fact that we also have a responsibility to expand the nature of knowledge within the broader field of research. It is different than hiring a consultant who's going to come and just do your work for you. This is going to be a partnership in which both sets of needs must be able to be met.

>> **Presenter 2** I think there are three key issues to consider. First, I think it is important to map out the policy or legislation barriers to sharing data and identifying when those barriers are real and when they are just barriers that we learned or handed down across the ages but that are not really part of the policy or legislative language—policies like Family Educational Rights and Privacy Act (FERPA) and the Health Insurance Portability and Accountability Act (HIPAA), and who is covered or not, or who is covered under which policy. And then if there is an actual policy barrier, we need to know where it is, and know how it needs to be addressed.

Second, we have been talking a lot about "systems" at the conference. I think we need to be clear about what a system is and what a system isn't. I think there is a need to map out what your particular child welfare system is and who the partners are in that system, because that has implications for data sharing and what information we need to feedback into that network.

Third, I have seen time and again how important it is not just to have case managers and frontline workers at the table in designing projects, but also when discussing systematic ways to share data back. It is a consistent frustration for caseworkers who are faced with entering data that goes to a state or a county but never gets back to them, so they don't see outcomes of their cases. That is on us researchers to make sure that that data gets back to the people who are doing the work so that they can see it as value in their day-to-day one-on-one work with clients, not just as sort of this overall huge systems change.

>> **Presenter 3** The goal of promoting greater use of integrated administrative data systems is not an easy one to achieve. I'm a realist, and I'm for finding opportunities. I mentioned federal opportunities to reimagine Statewide Automated Child Welfare System (SACWIS) data systems. Sites should look for opportunities that fit within their capacity and identify points where information sharing might be more valuable. Identify where there are benefits or where there are potential partnerships to get information so you can make informed decisions and improve practice.

>> **Presenter 4** A barrier to this goal is that a lot of our communities and families are not comfortable with a high level of data integration. They are not comfortable with their doctor knowing that they have a child welfare case open, and they are not comfortable with their child welfare worker knowing that they went to the doctor last month for whatever it was that they went for. So that means that we have to bring our community along with us for this integration to happen.

>> **Presenter 5** I want to refine our language a bit. Integrating data is not that hard. It takes a lot of time, it takes a lot of money, and it takes a lot of work—but it is also extremely valuable. Let's just do it, it is a matter of political will.

To generate that will we need to develop ways to maximize return on investment for this work. As agencies become more evidence-dependent, they consume more and more information. New types of information become more expensive to produce because agencies have produced all the "easy stuff." Universities have the infrastructure to produce new information more efficiently. The relationship should be, "How do I help university collaborators produce the new information and then cycle back this information, or the methods they use to produce this new information, more efficiently for the public agencies?" That's a productive relationship between the universities and a healthy collaboration between universities and public and private sector agencies.

>> **Presenter 6** I think it is critical to set up a regular opportunity for "learning exchanges" between university partners and agency representatives. We do them quarterly. We identify a particular topic of interest to our agency partner, such as a topic they want to know more about or that relates to a policy or practice change they are considering. The researchers spend about 45 minutes highlighting key pieces of information or key things we don't know, and then we spend the next 2 hours talking to program administrators about what the information means for

programs, practices, or policies within the agency. The other thing we do is attend regular briefings to agency leadership from key programs or divisions. These strategies ensure we have consistent, formalized mechanisms to keep these relationships going and that we are continuously implementing or exchanging information over time.

>> **Presenter 7** This conference has made me think a lot about what really strong research partnerships look like, and how to support the formation of strong partnerships at the outset of a collaboration and maintain them over time. There is also the consideration of building capacity for this type of work. I think it is important to continue these types of conversations, both in-person like we have been doing at this conference and by providing information in non-academic products and forums to convey all the complex work that everyone does so that it is distilled to key points that communicate findings, as well as lessons learned and challenges with regard to process. We don't discuss the importance of dissemination enough, or of considering multiple avenues for getting our work out there once it's completed. That goes beyond meeting with your partners and reporting back results or of publishing findings in scientific journals. It also means translating findings from work in one setting to other jurisdictions.

>> **MOD** How can the child welfare field promote greater use of integrated administrative data? Is it possible to have a national agenda on this topic, or must this effort be focused on local needs?

>> **Presenter 5** We need to create one national database that allows you to track people from the time they enrolled in the child welfare system to the time they leave, and the things that happens in between, and the database should only include about 10-to-15 pieces of information. Once that system is in place, we can figure out how to use it and grow it organically from there.

>> **Presenter 8** I want to build on that suggestion. We have the National Center for Health Statistics, we have our National Vital Statistics system. In our state, every single year we create linked birth and infant death files. That linkage happens to data available to researchers. What if we started making requirements that all the Child Welfare data goes to this entity and gets linked to birth data, and we go from there?

>> **Presenter 1** To support the overall goal, we need to talk about the national incentives for creating such a system. At the federal level, they can provide financial support so that states can move rapidly to access technical assistance. The Federal government can also set statutes and provide guidance for these types of systems.

>> **Presenter 2** I also think that part of this initiative needs to be about sharing what we are learning with others through a range of outlets. The majority of people in the field do not read journals, and even those of us who do have trouble keeping

track of articles as they come out in various disciplines and across various areas. There should be a "learning repository" to communicate key issues like the types of methodological issues we face or ideas that work well, and so on.

>> **Presenter 3** Federal agencies are providing seed money to explore innovations in the use of integrated data that offer a great opportunity to "think outside the box" and look to innovative sites to learn from, or to do some of that education and sharing. I mentioned the Comprehensive Child Welfare Information System (CCWIS) work at a federal level. It provides funding opportunities to look at bringing in new data elements that haven't been brought into the child welfare data system to try to capture that complex issue of measuring well-being—maybe with education, or health, or behavioral health indicators. We need to know more about the challenges of bringing that data into child welfare.

There are also opportunities at the federal level to look at the issue of context—states are different, counties are different, their systems are different, their definitions are different. One of our projects with the federal government is trying to document these differences, so that we understand how state-to-state differences influence what are we seeing in variation in databases like the Adoption and Foster Care Reporting System (AFCARS) and the National Child Abuse and Neglect Data System (NCANDS) and some of the policy influences on these differences. Hopefully, the next few years will provide us more information along that line to look at these issues at a national scale to advance this field and link that data with other data sources.

>> **Presenter 1** What I really want, more than action steps, is a clear interpretation of policies like FERPA and HIPPA. In my reading of these acts, it seems clear to me that agencies can use these types of data for planning and evaluation purposes, but there are different interpretations of these rules. Clarifying what is permitted with respect to use of these types of data at a Federal level would be a really important contribution to moving such an agenda forward.

>> **Presenter 9** I want to push for two forms of data that I think we need. Those in the child welfare system wants to know whether what the system is doing "works" for children and families, but in order to know if it works we need to know what the child welfare system is actually doing. We don't actually know anything in detail about services, from a national perspective. Currently, state data reflects whether families receive a service or not, and maybe it reflects different types of service categories. We have no idea about dosage or length of service, or whether it's an evidence-based service. Second, we need better data about foster care providers and how they contribute to outcomes.

>> **Audience Member** Service-related data may not be stored in the structured data in the administrative databases maintained by child welfare, but that doesn't mean that we couldn't get access to them. I work in a healthcare system, and treatment data is stored in our electronic health record. We can use data to ask

questions about how youth are doing. In addition, caseworker notes may be available, but usually not in a format that is easy to work with—they are often qualitative and free text. Caseworkers are aware of how placements are doing, or when placements are about to break up. We haven't put all the tools in place to capture that information effectively. If we could put the infrastructure in place to access these data for research and analytic purposes, that would be critical.

>> **Presenter 2** My perspective is, what outcomes are you trying to achieve and what data do you need to evaluate whether you are achieving those outcomes? We use words like "well-being," or even "permanency" without being clear about what we really mean. We need to operationalize these concepts and be clear about what we are trying to do. Without that level of detail, it is hard to know what data I need to inform decisions.

>> **MOD** How do we come together as a field to operationalize a key set of outcomes that we could track—or that we need to track—across our systems? What outcomes would be responsive to the needs of local systems and communities, but also unifying enough that we could develop a national agenda around their use?

>> **Presenter 2** A complicating factor is that different systems share the responsibility for well-being. Some indicators may be the responsibility of child welfare, and others may be the responsibility of other system, or of multiple systems working in collaboration. If we set up well-being outcomes for child welfare that aren't under child welfare's control, it makes them dependent on external responses and externalities. We need to start organizing where one service stops and where another one begins. In some ways that is an argument for integrated data, but I also think it is an important consideration in this conversation.

>> **Presenter 5** For the most part, in terms of child welfare, it's a settled matter—our priorities are safety and permanency. If we want to expand on the things that we want the child welfare system to be concerned about, we are going up against a fairly substantial body of public policy that is designed to constrain state intervention in families. This analysis does take place in a policy framework, and for the most part, safety and permanency are what our government has figured out the child welfare system can do in response to poor parenting. If kids aren't safe, the state has the right to remove them from their parents for a period of time and put them under the care of another individual. Then the state has to solve the issue, and they need to do it quickly. That's sort of the scope of the business. Let's figure out how to run a better system constructed around safety and permanency. Once we figure that out, then let's go on and do the other things. But we also need to understand that adding those additional outcomes—like improving educational outcomes—may negatively affect our focus on things like permanency.

>> **Presenter 9** I want to disagree, a bit. Safety and permanency are the primary goals of the child welfare system, because we think they are different and should

promote well-being. I would say that they don't, necessarily—at least with regard to permanency, we are not sure that is true. Second, that approach assumes that these goals work together. We think of foster care, in some ways, as a solution to problem of child safety—that's why we remove kids—but it also creates a problem for permanency. When you have a solution that's also a problem, there is not a simple answer.

>> **Presenter 1** I also want to chime in here and think about it differently. I think figuring out what the ultimate outcome that we are looking for is really important, and at some point, we are going to have to measure that. I want to draw the parallel to what the folks in child well-being in the education system have done in order to look at things from a system-level perspective and from a community-level perspective. Education system outcomes revolve around five key indicators that I think of as "dominoes." The last one is whether kids go on to college. That is really important for well-being, but that's the last one. It starts with whether kids show up to school ready-to-learn. The education system looks at children's kindergarten readiness assessment, they look at their third-grade reading scores, they look at sixth grade attendance, and they look at high school completion, and then they look at college matriculation. All of those "outcomes" are a domino to one another, and if a child is failing on one of the early indicators, we know that the other ones are not going to happen later. Now, my question is, what are those key indicators in the child welfare system that could tell us whether our kids are going to have a healthy outcome down the road, and can we measure these in our systems? We have to find the outcomes that we can easily measure that act like early indicators that ultimate outcomes, down the road, are on track or are not going to be what we want to see.

>> **MOD** Just to bring the conversation back to the issue of integrated data systems and how to move that part of the agenda forward—we have heard a couple of potential models to consider. We heard the idea of identifying a set of indicators that could be tracked longitudinally across a wide range of settings; we heard the idea of linking together some of the federal data with birth and death records. We've discussed some federal initiatives to support this kind of work. We have our federal partners at the table here, as well—what could we do to promote more efforts to support an integrated data system for our field that supports efforts to better understand issues affecting child safety and well-being.

>> **Presenter 8** I would like to frame that a bit differently—I would suggest we think about these efforts as producing population-level surveillance data, at a national level, that provides information about the characteristics of children identified by communities as having concerns about their safety, health, or well-being, broadly defined at this point.

>> **Presenter 5** Well, I think it's going to be important to understand that the party has already started. There are a lot of people working on these issues—state and local governments, colleges and universities, advocacy groups, Offices of

Management and Budget—you have a variety of players already in this space. It is important to ask this question in that context and bring these parties together to minimize duplication of effort. There is a tremendous amount of energy focused on this issue at the moment, and it is important to take stock of all of the work that's taking place that speaks directly to this question.

>> **Presenter 6** We should decouple this discussion from the child welfare system. Involvement with the child welfare system is one indicator that either services were not in place or services failed a child. They ended up involved in the child welfare system when we could have addressed a lot of problems through economic support, quality early childcare, or education, through other services. If we are really thinking about issue of surveillance—of what kids are getting or not getting, what kind of risks there are, etc.—I think we need to remove the child welfare system from the central focus.

>> **MOD** That brings us back to the suggestion that we work to leverage those public systems—like birth records or other vital statistics—as the basis for this work, rather than the child welfare system.

>> **Presenter 1** What is the benefit of this type of monitoring system for different constituents? I like to be able to look at certain indicators, but at a state level, I would like to hear why this type of system would be beneficial? How will these types of data result in better outcomes for kids at a local level?

>> **Presenter 8** In our state, we have a vital statistics advisory committee. The state sends these data and they get linked to child welfare data—it would become a part of such a national repository, but all of those data would go back to the state and could be requested and used by local public health departments, local researchers, etc. I am not suggesting that there be a national data system that would not allow that data to be used at the local level. I am just suggesting that such an approach strikes me as a useful and potentially achievable place to begin in terms of constructing a health surveillance system, or one more narrowly framed around child safety, defined by whether or not a child dies and whether or not a child becomes involved with the child welfare system.

>> **Presenter 2** One point that I don't want to lose is this notion of understanding what we are doing. One of the things that has been difficult to track is trying to understand how much to spend on the "front end" of the system versus other costs like administrative functions or foster care services, etc. As we are thinking about integrating various types of child welfare data, it would be helpful to make sure we include a financial component. This could allow us to ask questions about how policy changes influence costs or how system investments contribute to other outcomes. Not to pick on differential response, but in the literature those programs are very different. The initial programs were developed with an entirely separate set of resources outside the system that workers could use to provide materials

assistance to families. That is very different from what happens in other states that operate more like a two-track system but do not provide additional money or services for one track or the other. These kinds of questions can't be answered by our child welfare system unless we are also tracking that kind of financial data, as well.

>> **MOD** This has been a fantastic discussion. Let me briefly summarize some of the issues we've learned about at this conference over the past couple of days. We have learned a lot about the benefits and challenges of using administrative data for research on child safety and well-being, and—in particular—about the benefits and challenges of linking, or integrating, data from multiple separate systems. We learned about ways to leverage existing data systems, like Federal National Child Abuse and Neglect Data System (NCANDS) or Adoption and Foster Care Analysis and Reporting System (AFCARS) data sets as a building block for these types of analyses and identify risks of system contact or adverse outcomes. We learned about ways that states and communities might integrate other population-level data like vital statistics records to expand our understanding of risks for system involvement, or how states and communities might form multi-agency partnerships to create more dynamic integrated data systems. We also learned about ways systems might link in geospatial data to learn more about the unique needs and experiences of children and families who come into contact with the child welfare system.

Next, we heard a number of speakers address the ways in which integrated data systems might expand our understanding of service use or service outcomes. We saw a number of examples of such applications, including the foster care, housing sector, and health care sectors, and in other types of community-based child welfare services. These types of studies can contribute to the evidence base of real-world dissemination efforts and also expand our notion of "practice-based evidence" by studying what works in different child welfare agency locales, giving us an opportunity to take lessons learned from the field and test them out using administrative data from public systems like child welfare and related administrative sources.

Next, we had the opportunity to learn more about how these types of studies can be leveraged to inform public policy or foster system reform. We saw examples of this in the form of system change efforts that employ predictive risk modeling strategies, as well as a variety of other novel and innovative methods to guide practice and policy change. We also learned about the ways in which the Federal government is supporting efforts at data integration.

Finally, with these two panel discussions, we heard more about the ways that local communities are engaging in this type of work, or the needs they might have with respect to forming partnerships or communicating with other community agencies. We learned that relationships, trust, and reciprocity matter when trying to build academic partnerships with community partners to develop integrated data resources to support community-engaged research efforts, and that the needs of all partners should be considered in this process. We discussed the need to convene working teams early in the process to define project goals and methods. And we heard that such project teams should involve child welfare leadership and caseworkers, providers, agency Management and Information Systems specialists, academic researchers, and policymakers. We know that policy analysis and cost outcomes matter in decision-making, and we have learned the importance of integrating cost and finances into these types of research. And, finally, we learned that translation matters and that we need to do a better job of communicating what we are learning for these types of efforts.

Conference Overview

Child maltreatment is a significant public health problem resulting in substantial adverse consequences for children and families and for society at large. Individual costs are reflected in the psychological and physical suffering of victims. Societal costs associated with child maltreatment stemming from increased health care, criminal justice, and educational and economic burdens are estimated at $500 billion annually for the United States. Yet, federal, state, and local governments face substantial barriers in the identification and assessment of maltreatment and in providing intervention and treatment services. The scope and complexity of child maltreatment, coupled with the limited resources available to the child welfare system, underscore the need for programmatic and policy-level solutions that are demonstrably effective and financially efficient in promoting child safety, permanency, and well-being.

Over the past decade, the landscape for using data to inform child welfare system efforts has seen tremendous growth. Technological innovations have allowed for the accumulation and centralization of large datasets critical to identifying risks of child maltreatment and its negative consequences and to better target community and system response to these challenges. How can these data be leveraged to promote more effectives efforts to detect, prevent, and respond to child maltreatment? The purpose of this conference is to showcase emerging and innovative approaches in the acquisition and use of administrative data to inform the societal and governmental response to child maltreatment. This conference will highlight the use of multi-system data (or Integrative Data Systems) to conduct predictive analytics, risk monitoring, or policy and program-focused research and evaluation to inform child welfare system solutions.

Three sessions will cover the use of integrative datasets to predict the occurrence of child maltreatment, predict negative outcomes in maltreated youth, and target effective and efficient delivery of services. The conference will culminate in two panel discussions of collaborative data sharing, analytic approaches to predict

C. M. Connell, D. M. Crowley (eds.), *Strengthening Child Safety and Well-Being Through Integrated Data Solutions*, Child Maltreatment Solutions Network, https://doi.org/10.1007/978-3-031-36608-6

maltreatment and outcomes, and how these approaches can inform policy and program delivery.

Session I: Leveraging Administrative Data to Understand the Scope and Impact of Maltreatment

Moderator: Christian M. Connell, Pennsylvania State University

Lawrence Berger, University of Wisconsin: "Leveraging Harmonized Multi-System Administrative Data to Examine Experiences and Outcomes for Child Welfare-Involved Children, Youth, and Families"
Christopher Wildeman, Cornell University: "How the AFCARS and NCANDS Can Provide Insight into Linked Administrative Data"
Bridget Freisthler, Ohio State University: "Going Beyond Residential Neighborhood: Innovative Uses for Spatial Data using Linked Child Welfare Datasets"
Joseph Ryan, University of Michigan: "Administrative Data and State Partnerships: Findings from Education and Criminal Justice"
Emily Putnam-Hornstein, University of Southern California: "Assembling the Book of Life through Record Linkage"

Session II: Developing Integrated Data Solutions to Identify Effective Interventions for Child Welfare

Moderator: Sarah Font, Pennsylvania State University

Cynthia Osborne University of Texas-Austin: "Integrating Data to Foster Child Safety and Wellbeing"
Patrick Fowler, University of Washington-St Louis: "Housing Insecurity in Child Welfare: Designing Fair and Efficient Data-Driven Responses"
Elizabeth Weigensberg, Mathematica Policy Research: "Integrating child welfare and Medicaid data to identify and predict superutilization of services for kids in foster care"
Christian M. Connell, Pennsylvania State University: "Evaluating service effectiveness in real world contexts: Applications with statewide administrative data systems"
Melissa Jonson-Reid, University of Washington-St Louis: "Too much, too little, or just right? How integrated data helps identify impact and opportunity"

Session III: Informing Public Policy and System Reform with Integrated Data

Moderator: Max Crowley, Pennsylvania State University

Rhema Vaithianathan, Auckland University of Technology: "Challenges in the use of Predictive Risk Models for Child Welfare"

Christine Fortunato & Jenessa Malin, Office of Planning, Research & Evaluation (OPRE), Administration for Children & Families: "Leveraging Linked Administrative Data to Inform Child Well-being: Projects from the ACF OPRE"

Sarah Font, Pennsylvania State University: "Examples of using administrative data for policy-relevant research on the child welfare system"

Ramesh Raghavan, Rutgers University: "Constructing the Next Generation of Integrated Data Systems"

Fred Wulczyn, Chapin Hall, University of Chicago: "Is the Emperor Wearing Any Clothes? Commentary on Big Data, Big Science, and the Evolution of Integrated Administrative Data"

Session IV: Priorities and Action Steps for National Coordination

Moderators: Jennie Noll, Pennsylvania State University & Christian M. Connell, Pennsylvania State University

Part I (Noll): State child welfare administrators and others working in child welfare and allied fields will pose questions or share insights based on their work in state and county systems across the country.

Part II (Connell): Session and integrative speakers will convene in front of the conference audience to field questions, facilitate conversation, and discuss the possibility of a national integrative network.

Index

C. M. Connell, D. M. Crowley (eds.), *Strengthening Child Safety and Well-Being Through Integrated Data Solutions*, Child Maltreatment Solutions Network, https://doi.org/10.1007/978-3-031-36608-6